AF624014

Concept Publishing Co. (P) Ltd.
₹700/-
SINCE - 1974

STATISTICAL METHODS IN PSYCHOLOGICAL AND EDUCATIONAL RESEARCH

By the same author

1. Contemporary Fields of Psychology and Experiments
2. Modern Psychophysical and Scaling Methods and Experimentation

---

About the Author

Dr. M. Rajamanickam, formerly Dean, Faculty of Education and Professor and Chairman, Department of Psychology, Annamalai University, has been teaching Psychology to the Postgraduate students for about three decades. A 1966 Ph.D. of the Annamalai University, he has spent much of his time in research and specializing attitude measurements and kept himself actively in guiding research students and produced quite a number of M.Phil and Ph.D. Degrees. He has published a number of papers in National and International Journals. He is a prolific writer, has published a number of books on psychological themes both in English and Tamil. His books, Psychology-I (1977), Psychology-II (1978) in English and Tamil were used as text-books at Higher Secondary levels for more than a decade in Tamil Nadu, and his Fields of Psychology, Vol-1 (1973), Fields of Psychology, Vol-II (1974), and Physiological Psychology, (1983) were used as textbooks at Graduate and Postgraduate levels.

He has designed and developed several attitude scales, like the religious Attitude Scale, Reactionism-Radicalism Scale, Environmental Pollution Attitude Scale, Noise Attitude Scale, Air Pollution Attitude Scale, Temperature Attitude Scale, Ecological Attitude and Cognitive Scale and Family Planning and Birth Control Attitude Scale, etc.

He has been associated with a number of professional organizations and presented papers and also presided over several sessions at National Conferences. In 1980 he has started the Community Guidance Service Centre at Annamalai University, which is first of its kind among the Indian Universities and extended the psychological services to the rural folks. He is the Founder-Editor of the "Journal of Community Guidance and Research," founded in 1984 with the objectives of promoting research and encouraging young scientists in the field of psychology and other related disciplines. Many young scientists of various disciplines have built up their academic and professional standings through this Journal and they are very much benefited.

STATISTICAL METHODS IN PSYCHOLOGICAL AND EDUCATIONAL RESEARCH

by

Dr. M. RAJAMANICKAM, M.A., M.LITT., Ph.D.

Formerly Dean, Faculty of Education and Professor
and Chairman, Department of Psychology,
Annamalai University

CONCEPT PUBLISHING COMPANY, NEW DELHI-110059

All rights reserved, No part of this work may be reproduced, stored in a retrieval system, or transmitted in any form or by any means electronic, mechanical, photocopying, micro-filming recording or otherwise, without the prior written permission of the copyright owner and the publishers.

ISBN : 81-7022-803-4 (PB)

First Published 2001

© Dr. M. Rajamanickam (*b.* 1925)

Published and Printed by
Ashok Kumar Mittal
Concept Publishing Company
A/15-16, Commercial Block, Mohan Garden
New Delhi-110059 (INDIA)

Phones : 5648039, 5649024
Fax : 091-11-5648053
Email : publishing@conceptpub.com.

PREFACE

Having taught a course of study for a number of times for more than two decades it is quite but natural for any teacher to have an urge to present the subject-matter in the form and style that is suitable for himself and also that could be essential for the students who pursue their studies empirically. Any research worker in psychology and education should know that his/her data as they come to him/her do not bear any meaning unless some kind of statistical treatment is given to them and explained the meaning of them clearly. This book is based upon and written on the critical views expressed by many of the university teachers who have been testing the works of many researchers of the Indian universities for a considerable period of time. It has been observed that many research workers in India do not go for advanced statistical treatment of their data, and even if any of them attempt, never take pains to explain the results meaningfully, so that the readers may repeat the same procedure in their research works. Many of our researchers generally go for the conventional statistical treatments and are satisfied only with the simple percentages, the mean, the 't' test and the chi square. In the application of statistical methods to the behavioural sciences, the dispersion scores, the various levels of significances, the estimation of errors and the variances that determine the behavioural patterns are often emphasised by the American and the continental scholars.

Further, while planning a research work, the statistical methods put more emphasis on the choice of population, the sample, the sample size, the method of data collection, particularly, the probability and also the randomness, so that the results obtained through such samples can be extended to the whole population or can be generalized. If this is not done properly, simply depending upon incidental or accidental and inadequate sampling, the results would be certainly spurious. In such cases the research worker alone can be satisfied with his/her work, but he/she is doing a great disservice to the discipline and puts the more enthusiastic posterity in great disappointment.

This treatise is an attempt to enlighten the Indian researchers with certain essential and suitable statistical methods that are required for any type of research work in behavioural sciences for treating the data in a more meaningful manner. The standard deviation is an essential element in a distribution and, therefore, the individual and the group scores should be reported in terms of the SD units (or standard scores) above or below the mean and the data should be analysed. All these aspects are well explained in this book.

The Duncan's Range Test is a later addition to the psychological measurement. This may be of immense use to the researchers in planning any type of research work. This may substitute the analysis of variance and the 't' test when the mean differences are smaller. Duncan's Range Test is explained and illustrated with a number of examples. The Factor Analysis, a highly sophisticated statistical technique is dealt with all the important aspects, right from formulation of hypothesis and the selection of the sample, through the techniques of factor analysis, steps involved in factor extraction, the rotation methods, steps in rotation, rotation of factors, the two dimensional and the three dimensional rotations, the interpretation of unrotated factors and rotated factors in a simple manner with illustrations in such a way that anybody may be tempted to apply the technique in his/her research work. Complicated mathematical equations as far as possible are simplified and in certain cases reduced to a bare minimum. Wherever a researcher uses the statistical methods in his/her work, the most important thing is the manner in which one interprets and explains his/her results meaningfully, so that others repeat the same research work.

I hope, that this book may serve as a valuable guide and also as a text to those who are aspiring to apply the statistical methods in various psychological and educational research works and also to the students of Home Science.

March, 2000
Chennai-40

M. RAJAMANICKAM

ACKNOWLEDGEMENTS

Messrs. Addison Wesley Longman Ltd., Edinburgh Gate, Harlow, Essex, U.K. for reproducing Tables IV and V of the Statistical Tables for Biological, Agricultural and Medical Research (0-582-44525-6) authored by Fisher and Yates.

Messrs. Prentice-Hall Inc., Upper Saddle River, N.Y., U.S.A. for reproducing Tables 9.2 and 9.6 for Duncan's Range Test from pages 245 and 255, from Experimental Psychology, 3/E, authored by McGuigan, © 1978.

Messrs. Free Press, Glencoe, Ill., U.S.A. 1955, 1-3 for reproducing A Table of Random Numbers from the Source, The Random Corporation, A Million Random Digits.

CONTENTS

LIST OF FIGURES

LIST OF TABLES

PART I

DESCRIPTIVE STATISTICS AND MEASUREMENT

INTRODUCTION

The scientific study of any discipline is always associated with a branch of mathematical science, called the statistics, and it is often applied to the data of the study for interpreting the results. Thus, statistics, just like the spice, add flavour to the studies and make the studies more meaningful and reliable. The researcher need not have any doubt about the results of his study in applying the statistical methods, if he had conducted his study systematically according to the statistical principles.

The descriptive statistics are the basic statistical measures which can be applied to any study with strong conviction. Whatever, advanced statistics one aspires to apply to his study, the descriptive statistics would tell him whether the data are fit for it or not. The investigator by using the descriptive statistics can ascertain the characteristics of his data or the information obtained in an investigation. The frequency distribution of the data may be used to draw a frequency curve or polygon to test the extensity of the dispersion of the scores. If the investigation was correct, then curve would be normal, otherwise, the curve would be lopsided or bipolar or irregular.

Further, the measures of mean, median, mode, and the deviations would reveal the various characteristics of the data of the experiment or investigation. If the polygon is having a small diversion from the normal curve, it can be skewed and adjusted. The correlation methods also describe the nature and the characteristics of the data, whether the data can be associated with the data of another variable. The researcher's further treatment of the data also depends upon these basic measures. In this way the descriptive statistics become an important scientific instrument of any research work.

1

THE ROLE OF STATISTICS IN MEASUREMENTS

Introduction

Psychologists, like all other scientists wanted to make their subject-matter more accurate and meaningful. The numerical scores obtained by psychological tests have no value unless they can be organised and compared in a systematic and meaningful way. Though their attempts are of very recent in origin, the efforts were very much fruitful and they have succeeded in their efforts. Events may occur naturally or they can be made to recur by manipulating certain materials. Events, natural or manipulated have certain variations in them. This variation is of great concern for any one who is interested in the appraisal of certain values. In every field, appraisal or measurement is made only by some sampling. In every sampling measurement the variation is a persisting characteristics and it is measured.

For example, in measuring intelligence, aptitudes, personality, attitudes, prejudices, reaction time, skin resistance there is variation in every sample irrespective of the type of individuals or the group. Physical measurements such as height, weight, chest, skull, length of arm, nose, angular separation show variations between any two individuals. Physiological measurements vary, Anatomical measurements vary, Geological measurements vary. So also measurements of botanist, chemist, physicist and agriculturist vary. There might be also relationship between variables. All these revelations of variations and relationships can be measured and statistics is often said to be the science which studies the variations and relationships.

The operation of measurement has been investigated by philosophers and scientists in their thinking. Therefore, the concept of measurement

is becoming more and more broader and more logical. But in any measurement numbers are assigned to events and subjects to make it meaningful. Guilford and Fruchter (1987) defined measurement as the assignment of numerals or numbers to objects and events according to logically acceptable rules. We are using numbers in the measurements because the number system is highly logical and subjected to verification. When events and objects are assigned with the same number we say there is similarity or they have identity. But all events and objects do not have similarity and identity and they vary in all respects. In such case we measure the variation between objects or events and among events and objects. We measure the extent of variation and also of one event or object from another and also the extent of relationship between one object and another. For all these the psychologists depend upon a knowledge of statistics. Statistical techniques are extremely important to many areas of psychology.

What is Statistics?

Statistics is defined as a branch of mathematical sciences which is concerned with the quantitative data themselves dealing with facts of state. Statistics is also stated as the science, mainly dealing with numerical data assembled and classified so as to present significant information. Statistics offers a means of describing observations in a more understandable way. For example, we have administered three tests to a group of 15 college students and the data obtained, consisted of the test scores. These scores as they are, do not make any meaning. That is, without some further treatment to the data it is not possible to get any picture of how the class as whole performed. How the particular individual responded in relation to one another? What types the inter-relationships exist among the scores of the three tests. Thus, there are a number of ways in which the statistical techniques can be useful for a researcher.

One of the processes of study in a small group research is the interaction analysis which was developed by Bales (1950). This has been proved useful in studying interaction among the members, committees or other decision-making groups. The relative productivity of individuals working alone, and in a group can be compared. In a problem solving situation, four or five individuals working independently with a single solution, agreed upon by the same number

of persons working together, can be compared between the quality of the group solution and the individual solution.

Research activities in most of the sciences enliven by statistical methods. Particularly, the psychologists and educationalists make their research results more interesting and meaningful by applying the statistical methods to their research data. Since they find their researches evolve accurate results through the statistical analysis, they make it a point to improve their knowledge in statistics and also try to make use of their skills in the application of statistical methods to psychological and educational researches. Particularly in the field of experimental and applied fields of psychology the statistical methods are applied to a greater extent. Modern researchers in the fields of psychology, education, sociology, home science, population studies and economics began to think and plan their researches in terms of statistics.

For proper design of an experiment, the researcher is depending upon a knowledge of quantitative statistical procedures. The treating of the experimental data precisely and meaningfully a statistical approach is of immense help to the researcher. Therefore, statistics have become a major tool of psychology as they are for all other sciences. The present day psychological literature can never be understood by a student of psychology without a fair knowledge in statistical techniques. In dealing with the scores obtained in an experiment we often speak of two terms such as statistics and data. Authors define statistics in different ways. First of all we must know that statistics is one of the branches of mathematics. According to Guilford (1987) statistics as a specialised branch of mathematics enumerates data and their relation to metric data and satisfy specific conditions. Statistics also deals with discrimination and classification. These two are basic psychological processes applied even in simple psychological phenomena. Therefore, statistics can be also defined as the specialized branch of science of mathematics concerned with the collection, discrimination, classification and interpretation of quantitative data and with application of probability theory to the analyses and estimation of population parameters. Psychologists who make use of statistics for their purpose speak of two types of statistics. Guilford (1958) states these two types, as descriptive statistics and inferential statistics.

The Descriptive Statistics

The descriptive statistics mainly are concerned with description of the data in mathematical terms. They are employed in data collection, organisation and description of the data. They help the scientist to evolve huge amount of information in simple and clear cut form. The descriptive statistics also enables the scientist to present the relationship that may exist between the various kinds of data that he collects. D'Amato (1970) states that the aim of descriptive statistics is to describe meaningful properties of the sample data in terms of mathematical indices like the mean and the frequency distributions, measures of central tendencies, measures of variability and relationships.

Generally, statistics are used to describe the experimental data and the results in terms of numbers. The numerical values, evolved through the application of statistics to some experimental or research data would tell us the nature of the variable. The numerical values in statistics are stated in terms of some concepts. For example, the arithmetic mean, median, mode and geometrical mean are used to indicate the position of the result. Instead of telling in terms of average, we state statistically, mean and so on.

In the statistical application to the data of some research o experimental results, the variability is one of the statistics often given importance. It is because all the individuals in a group do not perform a task in the same manner or a group of students do not have the same height, weight and chest measurements. Every one varies from the other to some extent. How much one varies from every other? Unless every one is measured according to a standard measure we cannot say anything about the variability. There must be a standard and how much every one varies from it. A mean is worked out and from the mean the variations drawn. This will show how much an individual deviate or scatter from the group. This statistic is called standard deviation. This is one of the most important statistics which would tell us the amount of variation in a group of individuals. Any psychological measurement without the statistics of the mean and the standard deviation, Standard Error of the mean is incomplete and ill conceived in its plan.

In psychological studies we always involve more than one variable and we wanted to bring out the type of relationship the variables have and it is always presumed that the independent variable is having

influence upon the dependent variable. The statistics that test the level of relationship are the correlation coefficients. No other statistics can tell us the level of closeness between the two scores or the variables than this. The correlation method of statistics can be applied to test the type of relationship between any independent and dependent variables which may be very much useful for predicting the functions between and among the variables.

In these days of scientific spirit and inventions of new determining variables it has become absolute necessary to know the importance and the application of the correlation method in all the physical, and social sciences. The functions of descriptive statistics are revealed in various forms. The basic statistics like the central tendencies, the mean, mode, median, the dispersion values like the deviation, standard deviation, standard error of the mean and the statistics that bring out the relationship between any two variables have become very essential in all the research works, not only from the academic point of view, but also from many professional point of view.

The Inferential Statistics

This statistics deals with the subject matter beyond sample data and infer characteristics of the population from which the sample is drawn. Once the scientist has organised and described his data, he is facing the problem of generalizing from it. At the final stage his data represent only a small sample of observations. He has to infer his ultimate goal only from the limited sample about the general population characteristics. To achieve his goal he can employ several statistical procedures which are called inferential statistics. This statistics facilitates the researcher to evolve a generalization from his data drawn from the general population and also to determine the extent of probability, to state to what extent his generalizations are correct. He can also furnish the quantitative estimate of the confidence he has in his results. Inferential statistics is sometimes also called sample statistics.

The inferential or sampling statistics would reveal the nature of the sampling. A sample is always a small portion of a large population. We always assume that the small sample drawn from the large population should represent in all its characteristics the large population. How do we know whether a single sample really represent the large population or not. Our study or measurement mainly depends upon

this. It is because we are going to predict the characteristics of the large sample only from the result obtained in the small sample. The sample also should not be drawn in a haphazard manner. It should be drawn systematically according to some standard procedure if we want to predict the characteristics of the large population. But it has to be tested. At this stage, the sampling statistics seek the help of the descriptive statistics. The error variance, the standard error would serve as an instrument that could detect and tell us how far the sampling could identify or differ from the large population in its characteristics. In other words these statistics, the standard error, would tell us how far the sample drawn from the large population differs in its characteristics from the large population. This is one of the important aspects of inferential or sampling statistics. The level of difference in the characteristics between the sample and the large population is only inferred. This is why, the statistics is called inferential statistics. The characteristics studied in sample is only limited. It does not bring out all the characteristics of a population. Here the purpose is to predict the characteristics of the large population. With results obtained from the sample we attribute the same characteristics to the large population. This means, we are generalizing the characteristics to the large population.

The generalization should be made very carefully since we are attributing the characteristics observed in small sample to the entire population. Therefore, the error variance is applied. The standard error or the dispersion value should be as smaller as possible between the sample values and the population values. Suppose, for example, if the dispersion value is higher, then we can assume that the sample did not predict the characteristics of the large sample. From this we can infer that there is some error in the sampling drawn from the population or the sample drawn was inadequate. This is how the inferential or sampling statistics serve the purpose in psychological, educational and in other social sciences.

Further, statistical analysis reduces large and small quantity of data to a few required norms. The data might have been collected by any one of the several research methods using appropriate instruments and sampling techniques. Statistical techniques serve as valuable tools for analysing the data and presenting them in precise terms in tables for easy understanding of the meaning of the data. This type of statistical analysis may help us for a better overall understanding of the data.

Researchers are quite familiar with the concept 'average' which describes the characteristic of the trait or the variable measured. This statistic is very much useful for comparing the level of property present in any number of groups and also the difference and relationship between them. Statistical methods thus are very much useful to quantify the descriptions of the variables. But we must go one step further as to see how far the descriptions of the traits like the differences and relationships are reliably significant. It is by repeating the same experiment on other groups with the same instrument, that we come to know the extent of difference in the averages, in the measures of variations or in dispersions, in relationships and in comparing with the measures obtained from the group previously studied. There are several statistical procedures for estimating the extent of differences and relationships.

Further, we know the result obtained from the sample. We do not know whether the result is attributable to the whole population. Therefore, as stated above, several samples are needed from the same population and the results should be compared to make any generalization through the test of significance. At what level the two groups differ? Whether the five groups included in the study are homogeneous or heterogeneous? If heterogeneous, what is level of significance? These processes of statistics, which can help us to assess the unknown information from the known information, are called inferential statistics.

Statistical methods are also used for accurate predictions. For example a group of men applied for some mechanical work and we do not know who can be proficient on the job. In such situations the aptitude test scores may help us to predict the level of proficiency of the group of men on the job. Who can perform the job better without causing damage to the machine or without involving in accidents? Similarly the scholastic-aptitude test scores and the school grades can be compared and the student's future grades can be accurately predicted by the statistical methods. A researcher should also bear in mind that he should plan his research in such a way that it should yield data suitable for the application of available statistical techniques and sufficient information can be inferred from it. Statistical methods belong to the area of applied mathematics, and behavioural sciences are very much benefited by making use of these methods.

But in the recent times, in the realm of psychological and educational research there is wide variation in the application of statistical methods. In psychological researches, researchers are not very keen in making use of sophisticated statistical techniques in their researches. It is found that some use very simple statistical treatment in their researches, while some others merely depend upon the statistics but collect insufficient data and omit essential analyses. Still others go for advanced statistical treatment but fail to interpret the meaning of their data. But it is true that certain areas of psychology are less dependent on statistics, though the area of psychological measurement mainly depends upon statistics for interpretation of the results.

Whatever, the omissions and commissions have taken place in the application of statistics, psychologists cannot deny that psychological research was given a new dimensionality and innovations by the utilization of statistics. The experimental and research results were made precise, accurate, simplified and communicable easily only by the application of statistical methods. Hypotheses are formulated and verified easily and it is preferred very much by the professional and academic men.

2

FREQUENCY DISTRIBUTIONS

Introduction

The importance of the application of statistical methods in psychological and educational research is felt very much when psychological measurement has become one of the areas of studies. The researcher collects enormous amount of data in his research work. He converts the responses into numbers by giving numerical values to them. In the next step he manipulates the data and classifies them and puts everything in tabular columns. It is always better, for all our purposes, to tabulate or classify individuals or scores into certain large groups or categories. The tabulation of the data may be in the form of frequency table or frequency distribution similar to one shown in Table 2.2. The frequency distribution is the basic for all other descriptive statistics if the researcher faces with mass of data. Only from the frequency distribution all other statistics are derived. Graphic representations like histogram and frequency polygon, etc. are all drawn from frequency distributions.

What is Frequency Distribution?

This is one of the descriptive statistics dealing with the data for evolving other statistics. A set of individual scores as given in Table 2.1 does not express any meaning. We know only the numerical value, whether one is large or smaller and nothing else. Whether one tends to be at either end of the range or at the middle. Whether the scores are distributed equally over the range or collected somewhere in the range. To clarify all these first of all the researcher should work out the data for a frequency distribution. Frequency distribution is an arrangement of scores in

column. If the experimental data are presented in a table, it is called frequency table. If the data are presented in graph it is called frequency graph. The frequency graphs may take various forms like bar graph, line graph, histogram and frequency polygon. These are all visual or figurative illustrations for the experimental data. Frequency distribution froms the basis for all these and for several measures of central tendencies, such as, mean, mode, median, and also measures of variability. Therefore, frequency distribution is an arrangement of a set of scores of a group of individuals. It is the sequence of intervals. After obtaining the scores from a group they must be arranged in a systematic manner. The conventional technique for organizing and arranging the data is called the frequency distribution. All the scores are placed from highest to lowest in a vertical column if scores are few and the frequency of its occurrence is recorded or tabled next to each score. If the range of score is very large it is desirable to condense the frequency distribution by grouping individual scores into intervals. The advantage of interval grouping is that it makes easy to tally scores. For examplc the following scores represent 52 students attitude towards religion on a summated rating scale. These scores as they are have no meaning.

Table 2.1 : Religious attitude scores of 52 students

64	61	84	88	64	60	56	63	68	84
68	50	74	54	57	69	67	77	59	70
63	86	70	72	67	66	59	72	87	79
76	71	94	82	81	67	75	68	75	90
68	73	74	58	66	63	73	76	64	88
83	68								

The table scores range from the highest 94 to the lowest 50. With these scores it will not be possible to have class interval of one since the range is very wide (94 – 50 = 44). Therefore, a range of 44 score unit is required to include all the scores. It is desirable to condense the frequency distribution by grouping the individual scores into intervals. Grouping the scores into classes simplifies the frequency distribution without sacrificing the accuracy. In this case since the range of scores is 44 it is better to have nine class intervals with five units or scores in each. It is always better to start with lowest scores and multiples of them for the size of class interval. It is also better to have the class size with an odd number scores like 3, 5, 7, 9 that the mid-point of the class will be a whole number without fraction. If the interval is with 3 then start with 6, 9, 12, etc. If it is 5, then start with

10, 15, etc. To have the class intervals, in this example, divide the range 44 by a suitable class size. Three seems to be too small. Seven is too large. Therefore, in between 3 and 7 there is 5. Dividing 44 (assume 45) by 5 we get 9 class intervals. On this basis the class intervals are set up and given in Table 2.2.

Table 2.2 : Showing the frequency distribution of 52 scores from Table 2.1

Class interval	*Scores/Tallies*	*Frequency*	*Deviation*
90-94	94, 90	2	+4
85-89	88, 86, 87, 88, 85	5	+3
80-84	84, 84, 82, 81, 83, 80	6	+2
75-79	77, 79, 76, 75, 75, 76, 78, 76	8	+1
70-74	74, 70, 70, 72, 71, 73, 74, 73, 72, 72	10	0
65-69	68, 65, 69, 67, 67, 66, 67, 68	8	–1
60-64	64, 64, 60, 63, 63, 63, 64	7	–2
55-59	56, 59, 59, 58	4	–3
50-54	50, 54	2	–4
		N 52	

There are nine class intervals and in each class five scores are included. Against each class interval the scores which fall within the class are recorded since the frequencies are not too many in each interval. Instead of entering the scores directly, tally marks could be used. But if the number of scores is not large entering the actual scores is convenient. The number of scores falling in each class interval is entered under the third column head, frequency. The frequencies can be totalled and verified whether there are 52 or not. In the frequency distribution column it can be observed that there is the usual tendency for the scores to be more concentrated toward the middle of the range and getting smaller and smaller at the two ends. In making use of the scores of class intervals it is to be noted the exact limit of the class intervals. For example in Table 2.2, in the class interval 50-54 the score of 50 is really from 49.5 to 50.5 and for 54, is from 53.5 to 54.5. The interval with scores of 50-54 covers the range of 49.5 to 54.5 in the measurement scales. This is the general principle followed, to start just 0.5 step below the ends. The purpose of the frequency distribution is to have a meaningful picture of the whole distribution of scores.

Histogram

A visual picture of a frequency distribution can be obtained by drawing a histogram to the distribution of scores given in Table 2.3. The histogram is also called column chart. This is done by placing the scores (grouped in the class intervals) along the horizontal axis (*X* axis or abscissa) and placing the frequency numbers on the vertical axis (*Y* axis or ordinate).

The line drawn horizontally from *Y* axis to the point on the graph on which the vertically drawn *X* axis line meets is called abscissa of the point. The vertically drawn line from *X* axis to the point on the graph at which it meets the line drawn from *Y* axis is called ordinate of the point. The two values, the abscissa (*X*) and ordinate (*Y*) are called coordinates. On the horizontal axis the midpoints 52, 59 etc. of the class intervals should be noted and not the limits of the intervals. The midpoints as single scores may be taken as the representatives of the scores included with the respective class intervals. Every midpoint has its corresponding frequencies and the midpoints and the frequencies are plotted on the graph. The midpoint and the frequencies are the coordinates. For example, in Table 2.2, 52 and 2, 57 and 4, etc. are coordinates. They are from columns 2 and 3 of Table 2.2. On plotting the frequencies against the corresponding midpoints and joining the points with vertical lines from *X* axis and by horizontal lines from *Y* axis we get rectangles of equal size, a single rectangle representing a single midpoint score. Each score in the distribution represents in the form of a small rectangle. From the histogram it is known that the scores are collected at the centre of the distribution, and getting thinner at the ends (Fig. 2.1).

The histogram is a series of vertical bars drawn to the frequencies with a common boundary line between two adjacent frequencies. The height of each bar represent one frequency. Each bar occupy certain area and the sum of area of all the bars is equivalent to the total number of cases or the whole area of the histogram. If we divide the histogram by the *N*, then there will be 52 small rectangles, one for each score. The area of each small column of the histogram can be expressed in proportion to the total area of the histogram by dividing each frequency of the class interval by 52. The proportion for each frequency is given in Table 2.3, column 5. The sum of the proportions of the small rectangles of the histogram should be 1.00. This is presented in Table 2.3.

A histogram to the same measures can be drawn to the proportion and there is no difference between histogram drawn to the frequencies and to the proportions. Both have the same areas. This is drawn to the scores given in Table 2.3 (Fig. 2.2).

Table 2.3 : Showing the frequency distribution of scores on religious attitude scale

1	2	3	4	5	6	7	8
Class Interval	*Midpoint of class interval*	*Frequency*	*Smoothed f scores*	*P of F* $\frac{f}{N}$	*Cumulative f scores*	*C.P.*	*Upper limits of intervals*
90-94	92	2	3	.04	52	1.00	94
85-89	87	5	4.3	.10	50	.96	89
80-84	82	6	6.3	.12	45	.87	84
75-79	77	8	8.0	.15	39	.75	79
70-74	72	10	8.7	.19	31	.60	74
65-69	67	8	8.3	.15	21	.40	69
60-64	62	7	6.3	.13	13	.25	64
55-59	57	4	4.3	.08	6	.12	59
50-54	52	2	2.7	.04	2	.04	54
		52	51.9	1.00			

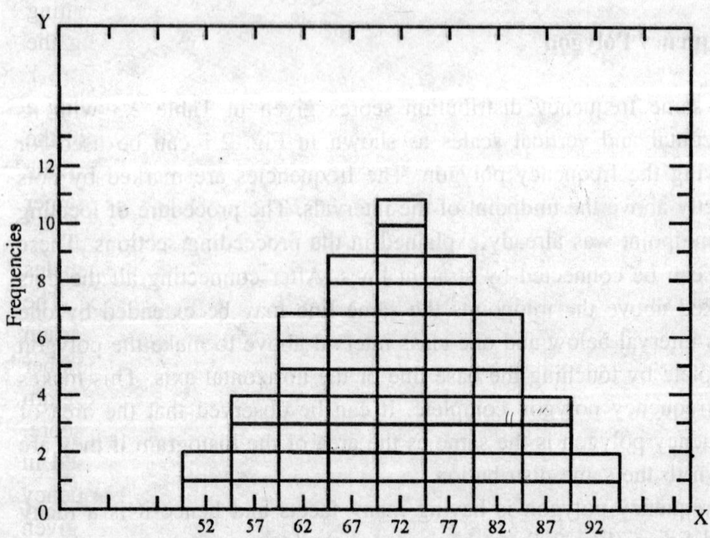

Fig. 2.1 : Histogram for the distribution of scores given in *Table 2.3* with frequencies and the corresponding areas.

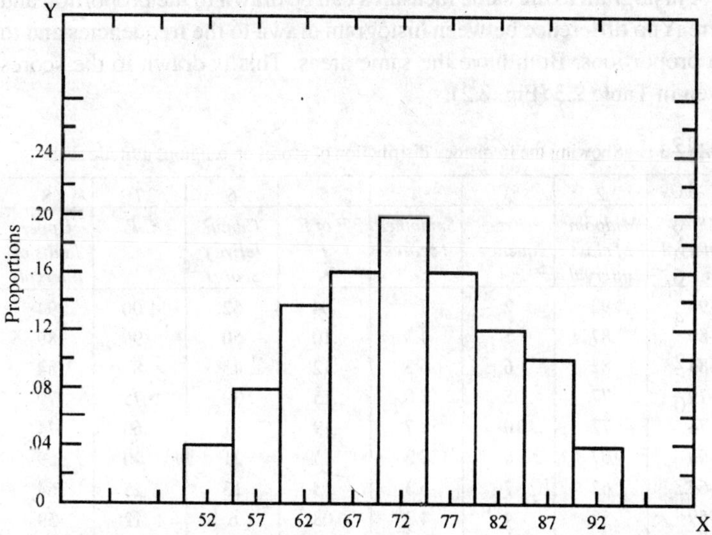

Fig. 2.2 : Histogram for the distribution of scores given in Table 2.3 to the proportions.

Frequency Polygon

The same frequency distribution scores given in Table 2.3 with its horizontal and vertical scales as shown in Fig. 2.1 can be used for drawing the frequency polygon. The frequencies are marked by dots directly above the midpoint of the intervals. The procedure of locating the midpoint was already explained in the proceeding sections. These dots can be connected by straight lines. After connecting all the dots marked above the midpoints the same line may be extended by one class interval below and one class interval above to make the polygon complete by touching the base line or the horizontal axis. This makes the frequency polygon complete. It can be observed that the area of frequency polygon is the same as the area of the histogram if they are drawn to the same distribution.

Frequency polygon is having many facets and hence it is a many sided figure (Fig. 2.3) not having smooth shape.

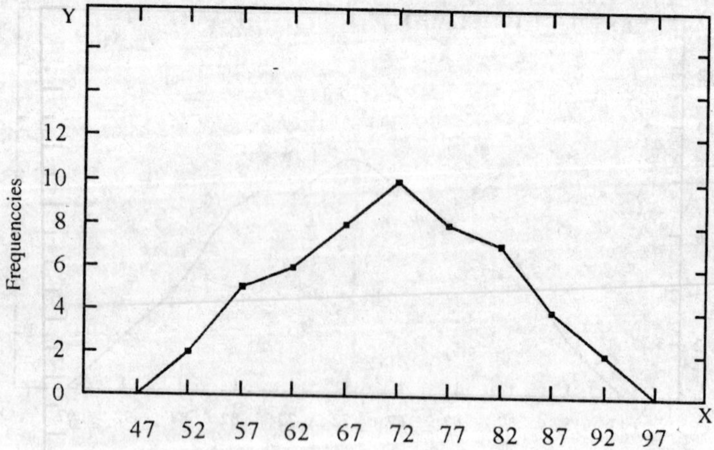

Fig. 2.3: Frequency Polygon for the distribution of scores given in *Table 2.3.*

Smoothed Frequency Curve

The histogram and the polygon show irregular shapes. This may be due to chance error. The irregular shape of the frequency polygon can be given certain regular shape by the process of smoothing the curve. It is by taking the averages over 3 intervals. The frequencies of the corresponding 3 intervals are summed up and divided by 3. The frequency of the interval to be smoothed and one frequency value above and one frequency value below are taken for summing up. The average goes for the central interval for smoothing. For example, in Table 2.2 the smoothed value for the interval 65-69 is worked out like this. The frequency value for the interval 65-69 is 3, and for the interval below 60-64, the frequency value is 7 and the frequency value for the interval 70-74 which is above is 10. By summing up we get 7+8+10 = 25 which is divided by 3 and we get 25÷3 = 8.3. The researcher can plot the original frequencies and the smoothed frequencies and can see the difference between the frequency polygon. This is illustrated in Fig. 2.4. The smoothed curve is called frequency curve, drawn to the scores in Table 2.3.

The frequency polygon is a curve which is superimposed upon the histogram if we disregard the vertical and horizontal lines. The areas of both the histogram and polygon are the same.

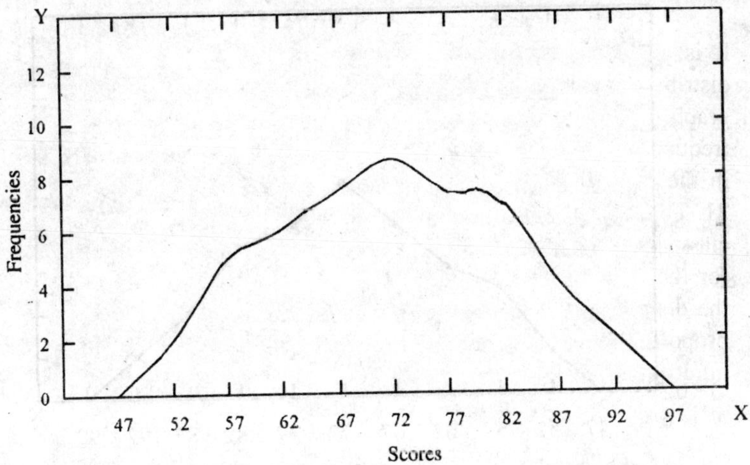

Fig. 2.4 : Frequency polygon for smoothed scores given in Table 2.3

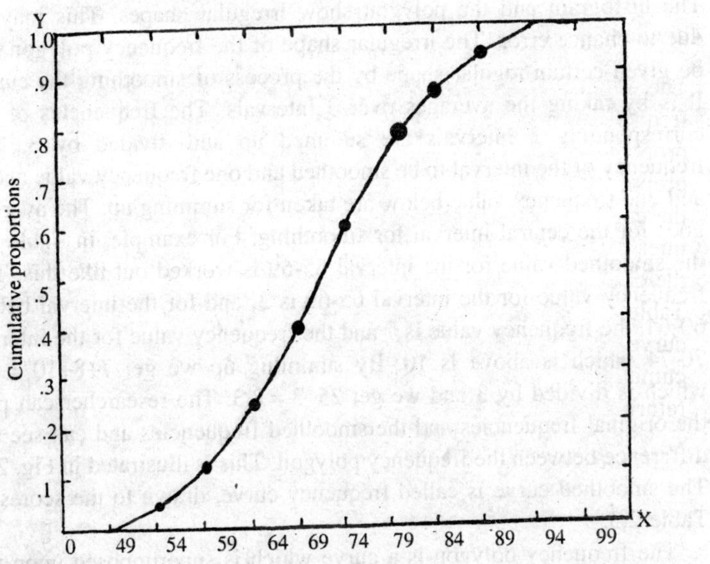

Fig. 2.5 : Cumulative proportion graph or ogive curve for the distribution of scores given in Table 2.3

Cumulative Proportion Graph or Ogive Curve

It is also possible to draw another graph from the same frequency distribution which is called cumulative frequency curve or percentage graph. The cumulative frequencies are obtained by adding successive frequencies from the lowest class interval to the highest. *For example*, in the illustration, the cumulative frequency for the interval 65-69 is 21. This value is obtained by adding the frequencies falling below the class interval of 64 - 69, 2+ 4 + 7 + 8 = 21. Thus the cumulative frequency for the highest class interval is 52. This is for all the frequencies in the distribution. This is given in Table 2.3. To obtain the cumulative proportions, we must divide the cumulative frequencies by the total number of cases N. It may be also obtained by another procedure. Since N is constant we can estimate the value of $\frac{1}{N}$ and multiply by the respective cumulative frequency.

In the illustration, for the interval 65-69 the cumulative frequency is 21. According to the first method, this can be divided by N to obtain the cumulative proportion $\frac{21}{52} = 0.40$. According to the other method divide 1 by 52 and multiply it by 21. Thus, we have $\left(\frac{1}{52}\right)21 = 0.40$. The same value we get. This is worked for all to the cumulative frequencies and presented in Table 2.3 column 7. A cumulative proportion graph can be drawn to the scores and to the cumulative proportions. This curve is also called as ogive curve. In plotting for the cumulative proportion curve we need not use the midpoints of the intervals as we have been using in other figures. The upper values of the class intervals are used for plotting this curve. This is presented in Table 2.3, column 8. The cumulative proportion graph or the ogive curve is drawn to the upper limit of the class intervals and to the cumulative proportions. The curve is like a learning curve of increasing return. (Fig. 2.5).

3

MEASURES OF CENTRAL TENDENCIES

Introduction

The frequency distribution of scores in the previous chapter, yielded for graphic illustrations such as histogram, frequency polygon to the original scores, frequency curve to the smoothed scores and the ogivc curve for the cumulative frequencies. It is observed that the curve is having a peak or the highest frequency near the middle of the curve of the *X* axis and gradually drops off on either side of the centre towards the base line or *X* axis. This means that a frequency distribution or frequency polygon or frequency curve has certain features. First of all we could see that the scores are clustering around the centre. Secondly scores moving away from the centre towards the ends. The curve either looks symmetrical or asymmetrical, either the curve is very steep or flat.

The distribution may yield a bell shaped curve if there was a systematic random sampling. All these illustrative descriptions do not fulfil the requirements of our scientific persuit. It is also important to know the value clustered at the centre and how far the scores scatter from the central value in the distribution. There are various measures like central values, dispersion, skewness and Kurtosis describe these aspects. A score at the middle of the distribution is considered as more representative than a score at either extreme and may provide more information.

The researcher, after having collected the scores from the sample, is in need of a statistics to present scores in an organized manner. They may be presented in various ways. The most commonly presented ways are the measures of central tendencies. A measure of central tendency is a number that characterizes the typical performance of the sample.

There are three most commonly applied measures of central tendencies. They are the arithmetical mean, the median and the mode. Each of these three measures has certain advantages and disadvantages. Therefore, what measure one uses depends upon the nature of the data that are being treated.

The Arithmetical Mean

This is one of the descriptive statistics used to represent the middle of a group. This is the common average of the measurement. Statisticians speak of all measures of central tendency as averages and one of them is identified as arithmatical mean. This is simply obtained by summing up all the scores and dividing them by the number of scores. The statistical formula to obtain the mean is $M = \frac{\Sigma X}{N}$, where M is the mean, Σ is the mathematical symbol for summing up the scores, X represents the raw scores and N is the total number of scores or cases. The formula is for estimating the mean for ungrouped data.

For the data given in Table 2.2, the arithmetical mean is 71.17. The sum of scores is 3701. This is divided by the frequency of scores or number of cases 52. It can be noted in Table 2.2 that no one scored 71.17. The arithmetical mean does not correspond to any actual score. It is an idealized performance of the group. If the group is fairly small this is the best way of getting the arithmetical mean. But if the group is considerably large this procedure is rather laborious. A better efficient computing procedures are available based on the class intervals and frequency distributions as given in Table 2.2. When the data are too lengthy for addition without the aid of a calculating machine it is better to apply another formula to compute the mean. The formula is arithmetical mean from grouped data. For the purpose of illustrating this formula we can reproduce the data given in Table 2.2.

The arithmetical mean was computed for the same scores using two formulas, the ungrouped data and the grouped data. There is very small negligible difference in the mean between the two formula.

According to the ungrouped formula the mean is 71.17 and according to the grouped formula the mean is 70.85 (Table 3.1). The difference is 0.32. When we apply group data formula for computing

the mean for large group of cases it may have small difference from the arithmetical mean.

Table 3.1 : Computation of arithmetical mean from grouped data

1 *Class interval* *Scores (Groups)*	2 *Xc* *Midpoint*	3 *f*	4 *fxc*
90-94	92	2	184
85-89	87	4	348
80-84	82	5	410
75-79	77	7	539
70-74	72	9	648
65-69	67	10	670
60-64	62	8	496
55-59	57	5	285
50-54	52	2	104
Sums		52 *N*	3684 Σfxc
Mean		$\frac{\Sigma fxc}{N}$	$\frac{3684}{52} = 70.85$

Where Σ is the symbol for sum of

f = the number of cases within the interval.

Xc = the midpoint of the class interval.

N = total number of cases.

Median

The central tendency of measure is better characterised by the median. The median is the most middle point of the sample. It is obtained by simply ordering the scores from lowest to highest and determining the score which falls at the middle point. The median may be defined as the midpoint of the scores of the measurement above which and below exactly 50 per cent of the sample or scores fall. This score is the same as the 50th percentile. The median is purely determined by the frequency of scores and any particular magnitude of score will not affect it.

The median is computed in the grouped data. The properties of median are the class intervals and the frequency distribution of the sample. This can be illustrated with the same data given in Table 2.2 and therefore, they are presented here in Table 3.2 and the median is computed.

Table 3.2 : Computation of the median with use of grouped data

Score interval	*Frequency*	*Cumulative Frequency*	
90-94	2	52	
85-89	4	50	18 = number of cases above the class
80-84	5	46	interval where the median is lying
75-79	7	41	
70-74	9	34	
65-69	10	25	25 = number of cases below the class
60-64	8	15	interval where the median is lying.
55-59	5	7	$Mdn = 69.5 + \frac{1}{9} \times 5 = 69.5 + 56 = 70.06$
50-54	2	2	
	N = 52		$Mdn = 74.5 - \frac{8}{9} \times 5 = 74.5 - 4.44 = 70.06$

In this measurement there are 52 cases and the median is the mid point on the scale above which there are 26 cases and below which 26 cases. The median lies somewhere within the class interval 70-74 and whose correct limits are 69.5 and 74.5. While counting the frequencies from the bottom case upward we can find that 2 + 5 + 8 + 10 + 9 = 34 cases. But we want only 26 cases. Since there are 34 cases to make the lower cases exactly 26 we need one out of 9 cases from the class interval 70-74. We find that 9 cases within this class interval are evenly spread over the area from 69.5 to 74.5. Within the range how far above 69.5 we need to go to include one case we need below the median. There are 9 cases in the class interval and we require one case. The total distance is 5 units. Therefore, on the scale measurement we go 1/9 of 5 or 0.56 units. Adding this 0.56 to the lower limit of the class interval we get 69.5 + 0.56 = 70.06 as the median. We can make a check whether this median is correct or not by computing the median using the upper measures. Counting the frequencies down from the top of distribution we get 2 + 4 + 5 + 7 = 18. We need 8 more cases from the next frequency of 9. We will have to go 8/9 of the way below the upper limit of the interval below 74.5. This works out 8/9 × 5 = 4.44. This has to be deducted from the upper limit. Then 74.5 – 4.44 = 70.06. This gives the median 70.06. This median obtained from the upper limit is the same as the one obtained from the lower limit. These two estimates must agree always. Otherwise there is something wrong in the computation.

The Mode

The mode is another measure of central tendency in a group of scores. A simple way of identifying the typical score is to pick out the score that occurs most frequently. It is the most frequently occurring score. Guilford (1975) defines that mode is the point on the scale of measurements with maximum frequency in a distribution. It may also be defined as the score value that occurs with highest frequency in a given set of scores. If the scores are few it is simple to pick out the score most often occurring. But if the group is very large then the scores are many and the simple method will not help us to compute it.

In such case the data should be grouped. The mode may fall somewhere near the centre of the distribution if the distribution is unimodal. Sometimes it may coincides with mean and the median. The mode may be also estimated from ungrouped data. There may be small difference between the modes estimated from ungrouped data and grouped data. Generally in a grouped data the mode is estimated at the mid point of the class interval which is having the highest frequency. To illustrate the estimation of mode the data given in Table 3.2 can be reproduced here.

Table 3.3 : Estimation of mode from grouped data

Score interval	*Frequency*	*Mid point*
90-94	2	92
85-89	4	87
80-84	5	82
75-79	7	77
70-74	9	72
65-69	10	67
60-64	8	62
55-59	5	57
50-54	2	52

A look at the frequency distribution in Table 3.3 will indicate to us the most frequently occurring *X* value. As the data shows the highest frequency is 10 which falls in the class interval 65 – 69. The mid point is 67 which is the estimate of the mode. Mode is often employed for quick inspection of the data to know the nature of the central tendency before we compute more complex statistics. If the frequency distribution is displayed

graphically the mode is the score corresponding to the highest point on the curve. The frequency polygon is a good graphic illustration for the mode.

Comparative Analysis of the Mean with Median and Mode

Authors often compare the three measures of central value, mean, median and mode to ascertain the accuracy in their measurements. The arithmetical mean and the median do not correspond exactly, but generally they do not differ very much. In our illustrative example the mean value is 70.85, the median value is 70.06 and mode is 67.0. The arithmetical mean is more preferred than the other two measures as it has several properties. It is preferred because it is the most reliable and accurate of the three central values. There is less fluctuation in the mean from sample to sample from the same population. The mean is also used for further arithmetical computations. Cases deviating from the mean (central value) provide important information about the distribution. Further, deviations are computed from the means. Guilford (1987) says that the mean is the centre of gravity of the distribution of scores.

Further, the deviations both the plus and minus, when summed up become always zero in the distribution only at the mean value. The median also may have this quality provided when it does not differ from the mean. But the median usually varies from the mean to a small extent. Another important characteristic of the mean value is that the sum of the deviations of the scores from the mean for ungrouped data become zero whereas in the cases of median, deviations do not become zero on totalling. Also while the mean deviations are squared and summed up, it is smaller than the median value.

The true mode is the greatest concentration in the distribution, the point at which more measures fall than any other point. The formula for estimating the mode roughly is:

$$\text{Mode} = 3\ \text{Median} - 2\ \text{Mean}.$$

The mode is crude and not a stable measure of central tendency. It gives only rough measure of the centre of the distribution. It is not often calculated with the same exactness as the mean or the median. But in the perfect normal distribution the mean, median and the mode will coincide. In a positive skew the median occurs on the left of the

mean. In a negative skew the median occurs on the right of the mean. In the former case the value of the median is less than that of the mean. In the latter case the mean is less than that of the median.

The central values, the mean, the median and the mode are used for the purpose of describing the scores. Among these three the mean and median are more important. While reducing the mass of scores to a few descriptive scores, the mean and the median are approximately the same and they are observed almost in the same typical manner, the average. Though the two scores are approximately in the same fashion the mean is having certain advantage over the median. The mean is supposed to be more stable than the median. For example, the means of two samples will be more closer than the medians of two samples. The mean may be further used for more advanced statistical treatment, whereas, the median cannot be used.

4

MEASURES OF VARIABILITY

Introduction

The central values may have the characteristics of clustering around the central values and scattering away from them. There are several measures to obtain the extend of clustering and scattering of the scores. Those measures differs from one another in interpretation and usage. Suppose, for example, there are two distributions *X* and *Y*. *X* has the scores as 9, 11 and 13 and *Y* has 3, 11 and 19. These two distributions have the same means and medians. But it is quite conspicuous that these two are different in some aspect.

The scores of the first distribution are very close to each other, whereas in the second distribution the distances between the scores are very wider. To state this process statistically, distribution *X* has less variability or dispersion than the distribution *Y*. This nature of variability is considered to be one of the most important characteristics of a distribution. The differences in variability are shown in the measures of central values as differences and these differences are due to the influence of the independent variable. That is, the influence of the independent variable is revealed in the differences of the central values. Therefore, what is variability and how it can be measured?

We have seen in the preceding section that the measures of central tendency provided us with information about the typical performance in a set of scores. But they do not indicate anything about the spread of individual scores within a distribution and how much they spread out from high to low scores. Suppose, for example a mental ability test was conducted on two groups of school students and whose average age was reported 14 and their average mental ability score was 110. From this we can conclude that both are of the same age group and

both are average in intelligence. Later on we are told that the age of the first group students ranged from 11 to 16 and of the second group from 10 to 17. The mental ability scores also were ranged from 98 to 110 for the first group and from 80 to 130 for the second group. From these informations we understand at once that there is definite difference between these two groups. There is *much dispersion in both the variables for the groups*. The first group seems to be more homogeneous than the second one in age and mental abilities. A measure of this spread is an important statistics for describing the groups. The way in which scores vary within a distribution is called *variability* and this may differ considerably even for groups that have similar means or medians. Therefore, *variability may be stated as the extent to which the scores in a distribution spread out from the average score or mean* in the group. To characterise a set of scores adequately it is important to have a measure of variability and also a measure of central tendency. The measurement *variability can be defined as the measure of scattered scores in a frequency distribution*. It is the measure of separate *scores spread around their central points*. The variability in also stated as dispersion and variation. The possible measures of variability are: (1) the range, (2) the quartile deviation (*Q*), (3) the average deviation (*AD*), and (4) the standard deviation (*SD*).

The Range

Of all the measures of variability, range is the simplest. Range is simple the difference between the highest and the lowest score. In other words, it is the *interval between the highest and the lowest scores*. It provides an estimation of the degree of spread within the distribution. However, the range depends only upon the distance between the two extreme scores in the total group. It does not tell us about how individual scores spread out within this distance. In our illustrative example the range is 94 – 50 = 44. It is not dependent and may change from sample to sample.

For example, a reasoning test conducted on two groups of students, consisted of 75 boys and 75 girls. The lowest score for boys is 85 and the highest 130. For the girls the lowest score is 75 and the highest 135. The range of the first group is 130 – 85 = 45 and for the second group is 135 – 75 = 60. From the results we can conclude that the *second group (girls) is more variable than the first group (boys) because the range difference is higher (60) for the second group.*

The Quartile Deviation

The quartile deviation is often called in statistics books as the semi-interquartile range and it is generally indicated by the letter Q. The semi-interquartile range is half of the scale distance between 25th percentile and 75th percentile. It is generally the middle cases of 50 per cent in a group fall between 25th and 75th percentiles. Just as the median divides the frequency distribution into two equal parts the quartiles divide a frequency distribution into four equal parts. The quartiles represent points on the variability scale. The first point is Q_1, the second Q_2 and the third Q_3. The second point Q_2 is nothing but the median of the central point. Therefore the formula for the quartile deviation is:

$$Q_2 \text{ or } Q = \frac{Q_3 - Q_1}{2}$$

Therefore, the distance between the quartiles Q_3 and Q_1 is called the interquartile range and the statistic that is stated as a measure of variability is the semi-interquartile range (Q). In the distribution of religious attitude scores we can locate the third and first quartiles by interpolation. One-fourth of the cases $\frac{N}{4} = \frac{52}{4} = 13$. Counting from the bottom to locate 13 cases, we find that we need 6 cases out of the 8 cases in the class interval of 60 – 64. As in the case of computation of median using the formula $\frac{6}{8} \times 5$ we get 3.75. After adding this with 59.5 + 3.75 = 63.25 is the position of Q_1. Similarly, counting down from the top we need 2 cases out of 7 cases in the class interval of 75 – 79. For the 2 cases we get $\frac{2}{7} \times 5 = 10/7 = 1.285$ and in deducting this from 75.5 of class interval we get 75.5 – 1.285 = 74.215 or 74.22 as the estimate of Q_3 (*Table 4.1*). Having estimated 63.25 as the Q_1 and 74.22 as the Q_3 we now must estimate the Q_2. This is given by $Q_3 - Q_1$ or 74.22 – 63.25 = 10.97. According to the formula $\frac{Q_3 - Q_1}{2}$ the semi-interquartile range is one half of this or 5.48.

Table 4.1 : Determination of Q_3, Q_1 and Q the semi interquartile range for the religious attitude scores

Class interval	*f*	
90-94	2	
85-89	4	
80-84	5	
75-79	7	Q_3 lies within this interval
70-74	9	
65-69	10	
60-64	8	Q_1 lies within this interval
55-59	5	
50-54	2	
	$N = 52$	

Average Deviation

In every distribution, scores deviate from the mean on either side to certain distance. The deviation is zero when the individual score is exactly the same with the mean. If the deviation is above the mean then it is taken as positive distance and if the deviation is below the mean it is taken as negative distance. To what extent every score deviates from the mean. This is estimated and the average of all the deviations is taken as the average deviation. In other words average deviation (*AD*) is the mean of all the separate scores. The formula to find out the average deviation is:

$$AD = \frac{\Sigma X}{N}$$

Where X = is the deviation of a score from the mean.

Σ = This means sum of. That is the total of all the deviations.

N = indicates the number of cases.

This formula is used when the scores are ungrouped. For example, in an experiment the following data are obtained from a group of 5 students:

$$5 + 6 + 7 + 8 + 9$$

The arithmetic mean can be computed by totalling all the five scores and dividing it by the number of cases. Applying the formula we get:

$$\frac{5+6+7+8+9}{5} = \frac{35}{5} = 7$$

This 7 is the arithmetic mean and we can find out the scores that are spread out around the mean score 7. Now we subtract the mean from each score to find out the difference. Thus, we have –2, –1, 0, +1, +2. These are the deviations of scores from the mean. For estimating the average deviation all the deviations should be summed up and divided by the number of cases 5:

$$\frac{(-2)+(-1)+0+(+1)+(+2)}{5} = \frac{(-3)+(+3)}{5} = \frac{0}{5} = 0$$

The average deviation in this case is 0 as the positive deviations equate with negative deviations.

The formula to estimate the average deviation from grouped data is:

$$AD = \frac{\Sigma fx}{N}$$

Where f = indicates frequency of cases
x = the deviation of a score from the mean
Σ = the product of x.
N = the number of cases.

But average deviation is rarely used in the measurement of variation in modern statistics. Some writers call it as mean deviation.

Standard Deviation

The most reliable measure of variability is the standard deviation, (*SD* or σ). This is most frequently used variability in the experiments and research works. This serves as an indicator of the degree of dispersion. The important information of any experimental studies or research works often rests upon only standard deviation and not on the mean. If different random samples are taken from the same group then the *SD* varies very little. Therefore, it is often said as the dependable and reliable measure of variability in the population from which the sample is drawn. Standard deviation cannot be explained in any simple terms with pictorial and geometrical illustration. The *SD* is nothing but representation of the distribution of scores. It increases proportionately with the increase of range. If the scores are spread out more widely the *SD* also increases proportionately.

There are two methods applied to compute the standard deviation. One is employed to ungrouped data and the other to grouped data. The formula for ungrouped data is, $SD = \sqrt{\frac{(X-\bar{X})^2}{n}}$, X is individual score, $\bar{X}$ the mean, Σ the sum of, n = number of cases. We can illustrate it with a simple example. Suppose, we have 5 scores of 5, 6, 7, 8 and 9 respectively. Adding these 5 scores together and dividing them by the number of cases we get the arithmetic mean $\frac{5+6+7+8+9}{5} = \frac{35}{5} = 7.0.$

Let us find out how widely these scores spread out around the mean 7. We can subtract 7.0 from each score. Then we have, −2, −1, 0, +1 and +2. These represent deviations of the scores from the mean. We will have to find out the overall measure of variability by estimating the average of these deviations. By summing up the five deviation values we get only zero (−2) + (−1) + 0 + 1 + 2 = 0. This means the minus and plus deviations around the mean balanced exactly. Therefore, we will have to do something else. One statistical way of dealing with the minus and plus signs is to square all the deviations. By this process all the minus signs become plus signs. The squared deviations are summed up and divided by the number of cases and the root of this value is computed to compensate the squaring of the deviations. The result of this is the standard deviation (*SD* or *S*). For the small sample of 5 cases the computation here follows applying the formula:

$$SD = \frac{(-2)^2 + (-1)^2 + (0)^2 + (+1)^2 + (+2)^2}{5}$$

$$= \frac{4+1+0+1+4}{5} = \frac{10}{5} = 2 = 1.414$$

We have calculated the average deviation (*AD*) elsewhere for the same five scores but the *AD* was found to be zero. The same 5 scores yield an *SD* of 1.414.

The standard deviation can be also computed from the grouped data or frequency distribution. The formula for the same is this:

$$SD \text{ or } \sigma = h\sqrt{\frac{\Sigma f(x)^2}{N} - \left(\frac{\Sigma fx}{N}\right)^2}$$

Where N = The total number of cases
f = Frequency
x = A deviation score
h = number of cases in each class interval
Σ = Sum of

To illustrate the computation of *SD* the data given in Table 4.1 may be reproduced here.

Table 4.2 : Computation of *SD* from grouped data

Class interval	*Frequency*	*x*	*fx*	fx^2
90-94	2	+4	+ 8	32
85-89	4	+3	+12	36
80-84	5	+2	+10	20
75-79	7	+1	+ 7	7
			(+37)	
70-74	9	0	0	0
65-69	10	−1	−10	10
60-64	8	−2	−16	32
55-59	5	−3	−15	45
50-54	2	−4	− 8	32
			(−49)	
	N = 52		$\Sigma fx = -12$	$\Sigma fx^2 = 214$

$$SD = 5\sqrt{\frac{214}{52} - \left(\frac{-12}{52}\right)^2} = 5\sqrt{4.12 - (-0.23)^2}$$

$$SD = 5\sqrt{4.12 - 0.053} = 5\sqrt{4.067} = 2.017 \times 5 = 10.085 \text{ or } 10.09$$

Presenting all the computations in summary from using the formula given above, we have:

$$SD = 5\sqrt{\frac{214}{52} - \left(\frac{-12}{52}\right)^2} = 10.085 = 10.09$$

Standard Deviation from Original Measurements

An alternative formula has been suggested by Guilford (1987) for computing the standard deviation without the knowledge of deviations. In both the methods (*SD* from ungrouped data and grouped data) we were in need of deviations and the means. But in this method we do not require either the mean of the measurement or the deviations. We

can straight away compute the *SD* from the original data. The formula is:

$$SD \text{ or } S = \frac{1}{N}\sqrt{N\Sigma X^2 - (\Sigma X)^2}$$

Where N = Number of cases or scores
X = The original score obtained from the measurement.
Σ = Summing up

Let us illustrate this method by an example. The scores obtained in a measurement are : 9, 11, 12, 10, 13, 14, 15, 16, 17. (*Table 4.3*)

Table 4.3 : Standard Deviation from Original Measurement

No	*X*	X^2
1.	9	81
2.	11	121
3.	12	144
4.	10	100
5.	13	169
6.	14	196
7.	15	225
8.	16	256
9.	17	289
	117	$1581 = X^2$

$$SD = \frac{1}{9}\sqrt{9(1581) - (117)^2}$$

$$= \frac{1}{9}\sqrt{14229 - 13689}$$

$$= \frac{1}{9}\sqrt{540}$$

$$= \frac{23.24}{9} = 2.58$$

The method of computation of *SD* can be used when the number of measurements are not quite large and values also are smaller in number and when a calculating machine is available and the accuracy also is not affected by this method. It can also be verified by the other method.

Interpretation of Standard Deviation

Standard deviation is the most valuable among the variability measures. The important characteristic of *SD* is that it would increase proportionately with the increase of the range. It mainly depends upon the range scores and is always measured from the mean of the distribution (grouped data). The standard deviation is fundamentally satisfied in its character. It is better to compare the standard deviation with the reference point, the mean. The *SD* is always associated with the respective mean

and mostly from the mean the *SD* is derived. If we want to compare the *variability of different groups* and measures *we use the coefficient of variation.* The formula for determining the coefficient variation is this:

$$V = \frac{100\, SD}{\text{Mean}}$$

The standard deviation is expressed as a percentage or ratio of the mean. The standard deviation has a very clear cut meaning for one particular type of distribution of scores. It is normal distribution. A normal distribution is an ideal frequency distribution defined by mathematical formula. It is represented by symmetrical, bell-shaped curve. In this, the scores are concentrated near the middle and tapering towards the two extremes. In the curve the heights of ordinates indicate the relative frequencies of each score in the distribution. The area under the curve over various score intervals indicate the proportion of the total number of scores fall in that interval.

The normal curve also illustrates the nature of mathematical relationship between the standard deviation and the proportion of cases. Thus in any normal curve about 68.2 per cent of cases fall between +1 and –1 standard deviations from the mean. About 95 per cent of cases fall between +2 and –2 standard deviations from the mean. Nearly all the cases (99.8 per cent) fall between +3 and –3 standard deviations from the mean. If an individual's score is one *SD* above the mean then he goes over 84 per cent of the cases. This means that he goes over 50 per cent of cases who fall below the mean and 34 per cent of cases who fall between the mean and plus one standard deviation.

In our illustrative example of the measurement of religious attitude among 52 university students the mean and the standard scores are computed as 70.85 and 10.09 respectively. Let us consider the scores of two individuals who scored 82 and 62 respectively and interpret them. The mean score of the group is 70.85 and the *SD* is 10.09. The first person with his score of 82 is 1 *SD* above the mean (70.85 + 10.09 = 80.94) mean score. The second person with his score 62 is 1 *SD* below the mean score (70.85 – 10.09 = 60.76).

The standard deviation is an essential characteristic of a distribution and, therefore, the individual test scores are often reported in *SD* units above or below the mean. This is done by *subtracting the individual score from the mean and dividing the differences by the standard deviation.* This result is known as *standard score.* Suppose a person

has scored 85, the mean of the group is 70.85 and the *SD* is 10.09. He has 85 – 70.85 = 14.15 points above the mean of the group. The 14.15 points by which *he goes above the mean is equal* to $\frac{14.15}{10.09} = 1.4$ *standard deviations.* That is, he is *1.4 standard deviation above the mean.* Another person has scored 95. He has 95 – 70.85 = 24.15 points above the mean of the group. This is equal to $\frac{24.15}{10.09} = 2.34$ standard deviations. This means he is *2.34 standard deviations above the mean.* A third person's score is 48.0. He has 48.0 – 70.85 = –22.85 points he goes below the mean of the group. This is equal to $\frac{22.85}{10.09} = 2.26$ standard deviations. *This person is 2.26 standard deviations below the mean of the group.* This is how the standard score is used to interpret the standard deviations.

The variability of scores within a set has been described in several ways. Four of them were described with some illustrations. Everyone has its own merits and demerits. For example, the range is the difference between the highest and lowest scores in the distribution. The range has two limitations. First of all the range depends upon the extreme values. The extreme scores may deviate from other scores more extensively and as a result of it the homogenity of the distribution is affected very much. Secondly the range as it is cannot be used as a scale for deciding anything.

The quartile range is another measure of dispersion widely spoken in psychological and educational measurements. Thorndike and Hagen (1970) state that the range is a better measure of variability. This measure of variability is statistically indicated as the semi-interquartile range, *Q*. Except that this will serve the purpose of indicating the high 25 per cent and the low 25 per cent of the cases in a data of large number, its other usefulness is very much limited.

The third index of dispersion we have discussed has been the average deviation (*AD*). This has become almost out of use at the present time as many resort to a better and advanced statistics, the Standard Deviation (*SD*). Edwards and Scannell (1968) state that this dispersion index of *SD* is quite stable and accurate since every score of the distribution is taken into account in estimating this.

5

THE IMPORTANCE OF THE MEASURES OF SKEWNESS AND KURTOSIS

Introduction

Researchers in the field of psychology and education are often resorting to conduct investigations for their researches in some population which they consider suitable for their studies. After having collected the data it is important that they should test whether their sampling or investigation was correct and the instruments they used for collecting data were accurate. Experts in the field of research often emphasise that this is an important aspect of data collection. The simple and traditional procedure of testing the sampling is to organize the data in a frequency distribution. It is better to condense the frequency distribution by grouping the individual scores into intervals. The interval grouping may give a clear picture of overall distribution of scores. The grouping of scores may be in intervals of three or five.

Frequency Polygon

A visual picture can be given to the frequency distribution by drawing a graph. This graph is called frequency polygon (Fig. 5.1). Sometimes it is better to arrange the range of scores by intervals than points. If this is done the graph is called histogram (Fig. 5.2).

We can see now whether the frequency polygon is normal or bell shaped curve. The idea behind testing the scores in this way is that in any investigation if the sample was drawn according to the random sampling technique and if the instrument was the correct measure of the trait then the scores should be spread in a bell shaped manner. There may be few extreme scores falling at the two tail ends and most

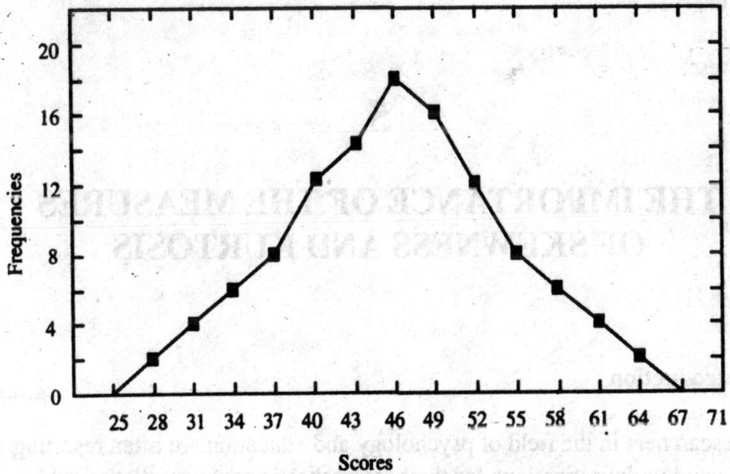

Fig. 5.1: Frequency polygon for the distribution of scores

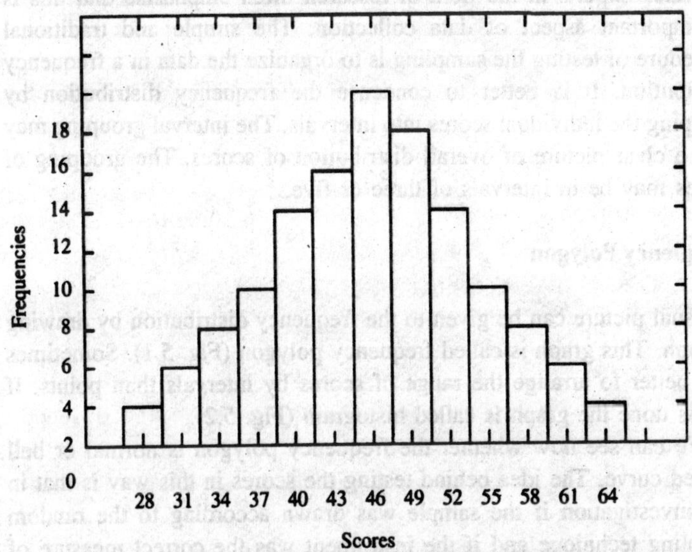

Fig. 5.2 : Histogram for the distribution of scores

of the scores should fall in and around the mean (centre) threrby making the curve a normal. This is what every researcher expects. This means that the shape of the frequency polygon indicates that the trait which the researcher measures is properly distributed in the sample of the general population. The curve is also called normal probability curve (Fig. 5.3).

Skewness and Kurtosis

Generally this occurs seldom perfectly in many investigations. This has to be described suitably. If we want to make precise and mathematical descriptions and comparisons the data of the frequency polygon should be converted into exact numbers rather than having them in graphic form. Statistic is the number that can enable us to deal with the data and help us to extract from it enormous amount of information. The statistic that can help us to describe the types of non-normal distributions are Skewness and Kurtosis. The two important types of non-normal distributions that are possible from poor sampling procedures are skewed distributions and multi-model distributions. If the curve is asymmetrical or non-normal we can describe the positions of the means and medians through Skewness and Kurtosis.

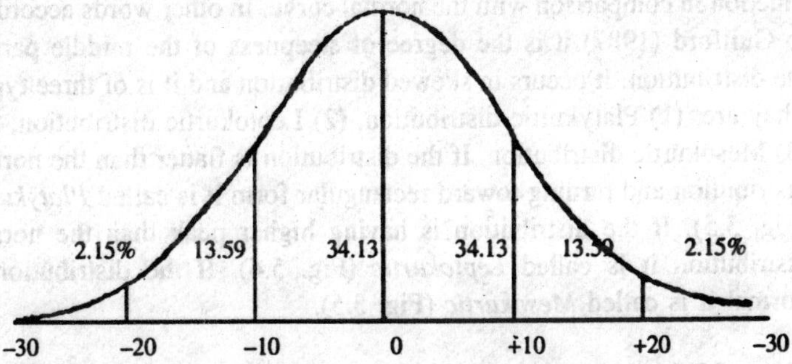

Fig. 5.3 : Proportion of area in a normal curve

In psychological researches if the distribution is found to be normal or symmetrical type then it can be described by mean scores and standard deviations and also we can proceed further. Instead, if the distribution is not normal or asymmetrical then the data cannot be used for descriptive purposes. It is because that the characteristic or the trait that is being measured is not distributed normally in the sample. For example, the distribution of income among the Indian families is not equal or normal and the educated people in semi-urban and rural areas are few and not proportionately normal compared to urban areas. If we don't have all these aspects in our mind and conduct an investigation and collect data the distribution would be non-normal or asymmetrical due to poor and unscientific sampling procedures. In such cases the sample scores may not represent every element of the population in proportion. Therefore, the sample is contaminated by the scores from some population other than the one the researcher wants to measure. At this stage we are in need of measuring the lack of symmetry. If the distribution is not having equal probabilities above and below the mean then the distribution is non-normal, distorted or biased. It can be observed that sometimes the sampling distributions particularly small samples may depart from normal distribution in various ways and some statistics reveal them. The statistics that shows the departures of the normal distribution of scores and the lack of symmetry is called skewness.

The skewed distribution may be a peaked type of distribution or a flat-topped type of distribution. This characteristics of skewed distribution is called Kurtosis.

Kurtosis means the sharpness of a peak on a curve of a density function in comparison with the normal curve. In other words according to Guilford (1987) it is the degree of steepness of the middle part of the distribution. It occurs in skewed distribution and it is of three types. They are: (1) Platykurtic distribution, (2) Leptokurtic distribution, and (3) Mesokurtic distribution. If the distribution is flatter than the normal distribution and turning toward rectangular form it is called *Platykurtic* (Fig. 5.5). If the distribution is having higher peak than the normal distribution it is called *Leptokurtic* (Fig. 5.4). If the distribution is normal it is called *Mesokurtic* (Fig. 5.5).

Skewed Distribution

A skewed distribution normally will have long tail or low end on one side and a short and high end on the other side. These two ends may occur on either side. A distribution in which the long tail of the distribution extends towards the low end or left is stated as *negatively skewed* (Fig. 5.6). A distribution in which the longer tail of the distribution extends toward high end or right is stated on *positively skewed* (Fig. 5.7).

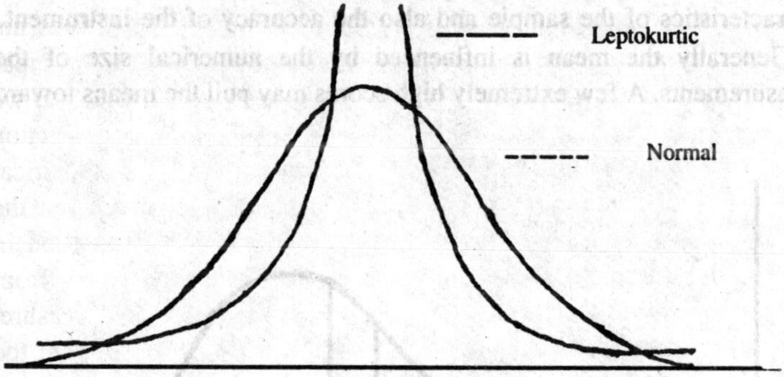

Fig. 5.4 : Leptokurtosis

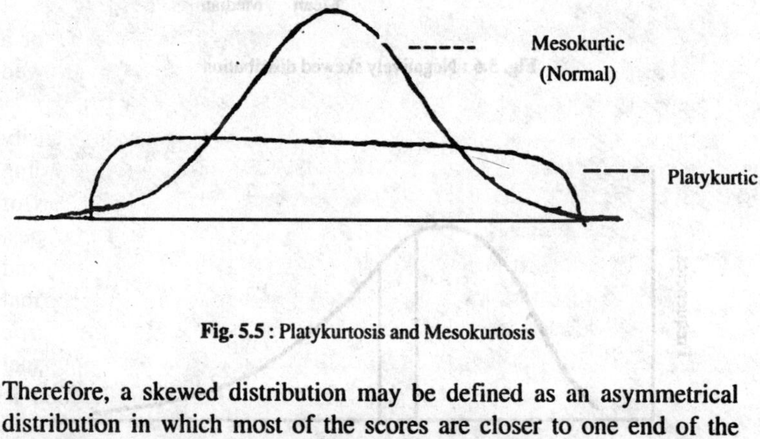

Fig. 5.5 : Platykurtosis and Mesokurtosis

Therefore, a skewed distribution may be defined as an asymmetrical distribution in which most of the scores are closer to one end of the distribution will have only one mode. The tail extremes are unequal in length.

If a researcher has obtained a skewed curve for the scores of his investigation, it is the indication that the sampling he has collected is of non-representative or the sample was not collected according to the random sampling technique or his instrument has been defective. The skewed distribution may occur if the instrument is defective in spite of correct sampling technique. Therefore, the skewed distribution is a departure from the normal distribution and it may indicate the characteristics of the sample and also the accuracy of the instrument.

Generally the mean is influenced by the numerical size of the measurements. A few extremely high scores may pull the means toward

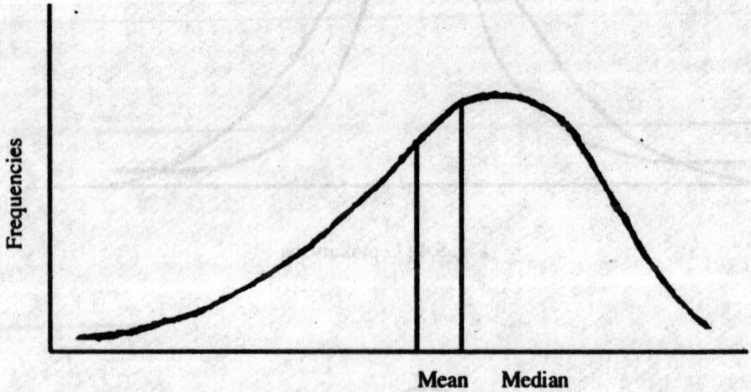

Fig. 5.6 : Negatively skewed distribution

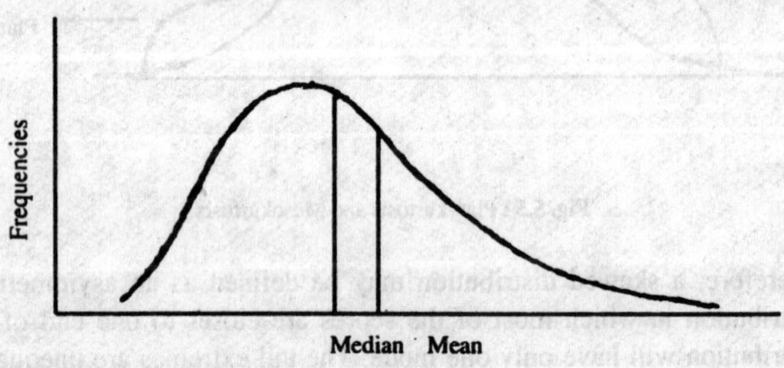

Fig. 5.7 : Positively skewed distribution

them and make tne mean to move from the centre of the distributions. Similarly a few extremely low scores may pull the mean toward them and make the mean to move toward the low end of the distribution. But the median is not affected by the numerical size of the extreme scores. The median will have always equal number of cases on either side. But the mean may be shifted to the left or to the right of the median depending upon the nature of the distribution. In a negatively skewed distribution the median is larger than the mean and in positively skewed distribution the median is smaller than the mean.

In computing the measures of skewness and Kurtosis the researcher should take a decision whether it is necessary or not and if he computes he must take into consideration the factors such as the nature of the research, the type of traits or characteristics studied, the size of the sample and the general population for the study. McNemar (1959) states that it is not advisable to compute these measures of dispersion if the size of the sample is less than one hundred. According to him symmetrical distributions in psychological studies often occur due to 'artifact'. That is, the shape is given by man. Similarly, the skewed distribution also is occurring in psychological studies due to artifact, a shape given by man. It is mainly because of the measuring instruments that such dispersions are taking place.

Suppose, for example a researcher measures the distribution of intelligence by using an intelligence test in a population. He included more number of industrial workers disproportionately in his sample of the general population. The industrial workers were educated only at the elementary school level. When the distribution of scores was made it has resulted in a positively skewed distribution (curve). Instead, if he would have added more number of graduates disproportionately in his sample of the general population then this would have resulted in a negatively skewed distribution (curve). Suppose, even if the sample were collected according to random sampling technique, having proportionate representation still he got positively skewed distribution (curve), the mistake is not with the sample, but with his test. His test was not easy enough for the industrial workers to answer the items and thereby differentiating the higher level of intelligence from the lower level. Since the distribution of scores yielded a skewed curve the researcher must re-examine his sample or his test.

We could see in a normal distribution that the curve is bell-shaped. Most of the scores fall in the middle of the curve and the curve is having a peak at one point. But sometimes, in some distributions the scores may be collected at more than one point unlike the normal one. If this kind of distribution occurred we call it a *multi-model curve*. In this distribution there are two peak points. It may be due to the nature of the sample or the characteristic of the instrument.

The multi-model curve is also known as bimodel curve. The sample consists of two elements of different types of the populations. The two peaks of the curves independently may have perfect normality. If the two peaks were due to sex or due to remote rural and urban populations, by limiting the sample either to male or to female or to extreme rural or to urban the curve may by perfectly normal. When a curve is drawn to the scores of a single investigation in which two populations are involved there would be two peaks of two normal curves. The curves contribute to multi-modality. Similarly, if the instrument is of bipolar type, having two types of scores there would be two curves which contributes to multi-modality. For example, if there is a neutral or 0 score in the middle and at either side are different weights of scores the curve drawn to the scores would be bi-model having two peak points. For example the Social Distance Scale developed by Bogardus (1933) has different type of scoring system. There are both positive scores and negative scores. When a curve is drawn to the scores of the individuals it will be multi-model type of curve.

The occurrence of skewness can be demonstrated by drawing a figure called frequency polygon and we can understand the nature of the dispersion. Skewness can be stated by a number of measures. One way of estimating the skewness is to find out the difference between the mean and median and dividing it by the respective standard deviation. This is the simplest way of measuring the skewness. The values required for this purpose are mean, median and the standard deviation.

It is to be noted that in a skewed distribution the mean is always drawn toward the skewed end of the curve. The mean maintains the centre of gravity of the distribution. In this process the mean is drawn toward the extreme scores. But, whatever, the position of the mean, the standard deviation maintains the equilibrium. The sum of deviations on one side of the mean is always equal to the sum of deviations on the other side of the mean. But the role of median in the distribution is that it only divides the area of the distribution into two equal parts, irrespective of the position of the mean anywhere in the distribution. In a symmetrical distribution the scores are distributed evenly on either side of the middle point. The mean, median and the mode all fall together on the same place. This is not the fit distribution for skewness.

6

CORRELATION METHODS

What is Correlation?

We have been concentrating in the preceding sections on the scores and the examples that vary along one dimension, mainly the measures of central tendency and measures of variability. They were described for the most part to characterise the distributions of single variables. But psychologist is also interested in something much more than these. In his attempt to describe and understand human behaviour, there are many situations and instances that demand for the type of relationship between the independent and dependent variables. The experimenter is interested in relating the value of dependent variable *Y*, to the independent variable *X*, whether the independent variable had any influencing effect upon the dependent variable. In many human problems the relationship between the independent variables and the dependent variables are very much involved and the psychologist is interested in describing the extent of relationship between the variables. One of the important tasks of a scientist is to analyse the interrelationship of the variables with which he is dealing with. Therefore, the psychologist is interested in knowing the degree of relationship between the level of aspiration and the actual achievement, between intelligence quotient and school achievement of the school students, the social psychologist is interested in the relationship between the socio-economic status and family planning and birth control of the general population. The industrial psychologists wish to find out the relationship between job satisfaction of the employees and their attitudes toward the management. Understanding such relationships are often found to be important for predicting one set of events to another. However, there is one statistic involving both the variables which can yield a single value, which is called the correlation coefficient. This has provided a very wide scope for the appraisal of psychological value. It is a matter of degree and we need some kind of statistical index to express the degree of relationship. The statistical index known as coefficient of correlation can express the degree of relationship and this can be computed. The coefficient of correlation is indicated by the symbol *r*. To make it more clear the often quoted example of the relationship between heights and weights can be mentioned here. People generally differ both in height and weight. For example, we can take any one group of people and find out their average heights and weights. We can also find out the variabilities of their heights and weights, also the correlation between their heights and weights. There may be certain inconsistency that may result in imperfect correlation. But there is a relation between these two measures in terms of degree which needs to be assessed.

If the correlation between two measures is perfect and positive it has a value of +1.00 and if there is no correlation its value is zero. If there is a negative and perfect correlation its value is indicated by –1.00. A correlation of +1.00 indicates that the group had highest score in one test and also had the highest score in the other test. A correlation of –1.00 indicates that the scores are in the reverse direction. It means the individual who scores the highest score in one test, scored the lowest score in the other test. A zero correlation means that there is no relationship at all. If changes in one variable tend to occur in accordance with changes in the other, the correlation between the two will fall some where between zero and 1.00. The correct value depends upon the extensity of their covariance. There are several measures of correlation in use in psychological and educational experiments and researches. They are the Pearson Product moment correlation coefficient, Spearman's coefficient of rank order correlation, Kendall's coefficient of rank correlation, Contingency coefficient of correlation, Phi-coefficient or four-fold point correlation, Point biserial correlation, Biserial correlation and the Tetrachoric correlation. Among them, the Product-moment correlation and the Spearman's rank order correlation are very often applied in Psychological and educational researches.

The Pearson Product Moment Correlation Coefficient Method

The valuable application of this method is that it utilises the values as they actually find in the distribution. But in the rank order method only the rank is considered and the values are not applied directly. The product moment method is used when the variables are in quantitative terms (*Table 6.1*). The basic formula applied to obtain the correlation value r_{xy} according to the product moment correlation method is:

$$r_{xy} = \frac{\Sigma XY}{N\,\sigma_x\,\sigma_y} \text{ or } \frac{\Sigma XY}{NS_x\,S_y}$$

r_{xy}	=	correlation between *x* and *y* variables
x	=	deviation of an *X* score from the mean of *X* scores
y	=	deviation of the *Y* score from the mean of *Y* scores
Σ_{xy}	=	Sum of all products of deviations of *x* and *y*.
S_x , S_y	=	Standard deviations of the sample distributions of *x* and *y* scores.
N	=	Number of cases
σ_x	=	Standard deviation of *X* scores
σ_y	=	Standard deviations of *Y* scores

Let us illustrate this method with an experimental data drawn from a group of 10 persons in tests *X* and *Y*.

X scores : 12, 13, 14, 11, 10, 9, 8, 9, 8, 7

Y scores : 12, 12, 12, 10, 9, 8, 7, 8, 7, 6

First of all arrange the scores in columns making sure of the corresponding *X* and *Y* scores are together in pairs. Secondly, compute the two means for *X* and *Y* scores as $\overline{X}$ and $\overline{Y}$. Thirdly, work out the deviation scores for *X* and *Y* scores as *x* and *y*, in two columns and check the sums which should be zero for both. Fourthly, square all the deviation and enter them in column 6 and 7. They are x^2 and y^2. Fifthly, sum up the squares of the two sets deviations. These are x^2 and y^2. Then compute the S_x and S_y (SDs) from these values. Sixthly, work out product for each person *x* and *y* deviations given in columns 3 and 4 and put them in column 7 and sum up them at the bottom. We get the sum of the *xy* deviations = Σxy. Then lastly the formula given above is applied to the data.

Table 6.1 : Product moment coefficient of correlation for two sets of scores obtained from 10 persons

S.No.	1 *X*	2 *Y*	3 *x*	4 *y*	5 x^2	6 y^2	7 *xy*
1	12	13	1.9	3.8	3.61	14.44	7.22
2	13	12	2.9	2.8	8.41	7.84	8.12
3	14	12	3.9	2.8	15.21	7.84	10.92
4	11	10	0.9	0.8	0.81	0.64	0.72
5	10	9	–0.1	–0.2	0.01	0.04	0.02
6	9	8	–1.1	–1.2	1.21	1.44	1.32
7	8	7	–2.1	–2.2	4.41	4.84	4.62
8	9	8	–1.1	–1.2	1.21	1.44	1.32
9	8	7	–2.1	–2.2	4.41	4.84	4.62
10	7	6	–3.1	–3.2	9.61	10.24	9.92
Sum	101	92	0.0	0.0	48.9 Σx^2	59.6 Σy^2	48.48 Σxy
Mean	10.1	9.2					

$$S_x = \sqrt{\frac{48.9}{10}} = \sqrt{4.89} = 2.21$$

$$S_y = \sqrt{\frac{59.6}{10}} = \sqrt{5.96} = 2.44$$

$$r_{sy} = \frac{\Sigma xy}{N\,S_x\,S_y} = \frac{48.48}{10(2.21)\,(2.44)} = \frac{48.48}{10(5.39)} = \frac{48.48}{53.0} = 0.899 \text{ or } 0.90$$

This method of product moment correlation can be applied if the number of cases are not too many. There is also an alternative short cut method for computing the Pearson Coefficient of correlation. This method does not require to compute the standard deviations of *x* and *y* or S_x and S_y. The formula is:

$$rxy = \frac{\Sigma xy}{(\Sigma x^2)\,(\Sigma y^2)}$$

This formula can be applied to the above data.

$$\frac{48.48}{(48.9)\,(59.6)} = \frac{48.48}{2914.44} = 0.898 \text{ or } 0.90$$

This formula yields the same result. The *r* is worked as 0.9. The calculation work is somewhat reduced and therefore, it is more economical from computation work is concerned.

In the examples given above we have been dealing with ungrouped data and with small number of observations. There is also another method of correlation for a Pearson r with ungrouped data. This method does not require the deviation scores or standard deviations. The correlation coefficient can be computed from the raw scores. The original scores can be straight away used for computation of correlation when the scores or cases are smaller in number. The formula is:

$$r_{xy} = \frac{N\Sigma XY - (\Sigma x)(\Sigma y)}{\sqrt{[N\Sigma x^2 - (\Sigma X)^2][N\Sigma Y^2 - (\Sigma Y)^2]}}$$

Where N = Number of scores or cases

X = Original or raw scores of X variable

Y = Original or raw scores or Y variable

If calculating machine is available for computation of the correlation coefficient it is the best method. We can illustrate the computation of the Pearson r with an example. It would be more convenient for calculation if the scores are in small numbers. The following are the raw scores obtained in two tests x and y from 10 persons.

X scores : 13, 14, 11, 10, 11, 9, 8, 9, 8, 7

Y scores : 14, 13, 12, 11, 9, 10, 9, 7, 7, 6

These two sets of scores are arranged in parallel columns pairing each other and squared in columns 3 and 4. The raw scores 1 and 2 are multiplied and given in column 5 and the formula applied to the data as shown in Table 6.2.

Table 6.2 : Correlation coefficient computed from the raw scores for ungrouped data

S.No.	*X*	*Y*	X^2	Y^2	*XY*
1	12	14	144	196	168
2	13	13	169	169	169
3	11	12	121	144	132
4	10	12	100	144	120
5	11	10	121	100	110
6	9	10	81	100	90
7	8	11	64	121	88
8	9	9	81	81	81
9	8	9	64	81	72
10	7	8	49	64	56
Sum	98	108	994	1200	1086
	ΣX	ΣY	ΣX^2	ΣY^2	ΣXY

$$r_{xy} = \frac{10\times1086-(98)(108)}{\sqrt{[10\times994-(98)^2][10\times1200-(108)^2]}}$$

$$= \frac{10860-10584}{\sqrt{(9940-9604)(12000-11664)}}$$

$$= \frac{276}{\sqrt{336\times336}} = \frac{276}{336} = 0.82$$

This is very simple method of computing the Pearson *r*. If a good calculating machine or computer available however large the sample is, this method can be used and there is no chance for committing any mistake.

The Scatter Diagram Method of Correlation

We have discussed so far that for small samples the Pearson correlation coefficient (*r*) can be computed by using any one of the methods described above for ungrouped data. But if the samples are large we must use the group data procedure by preparing the scatter diagram (Table 6.4). The scatter diagram procedure is somewhat shortest in computing the correlation. According to this method the first step is to find out the range of the scores between the highest and the lowest scores for both the variables and to select class intervals. The scores are grouped into convenient number of intervals. It is always better to have odd numbered intervals. A two dimensional tabulation sheet scatter diagram is prepared to indicate the class intervals of *X* scores and *Y* scores. The scores of *X* variables are shown on the top from left to right (on the ordinate) and the scores of *Y* variable from bottom to top on the left of the chart (on the abscissa). This is also called a two way frequency distribution. This method is considered to be very accurate for many purposes (Tables 6.4 and 6.5).

In the scatter diagram method of correlation, an individual's score on *X* variable is plotted against his score on the *Y* variable. The point where both the variables meet or intersect is indicated by a dot. In the same way for all the individuals the intersection points of *X* and *Y* variable scores are indicated by dots. The amount of dots in the cells is a direct indication of the degree of correlation between the two sets of measures. When there is no relationship the dots are distributed randomly. When the relationship is intense the dots tend to collect into an ellipse and the scattering become narrow. A perfect correlation between the measures makes the dots a straight line. This indicates that a change in one variable is also making change proportionately in the other variable. The direction of the line indicates whether the correlation is positive or negative. A positive perfect correlation will have the slanting line going from left bottom to the right top (Table 6.4), and a negative perfect correlation will have the slanting line going from right bottom to left top on the reverse direction of the positive correlation.

An Experimental Illustration for Scatter Diagram Correlation Method

Suppose for example, we have obtained two sets of scores from an investigation conducted to assess the relationship between the Religious attitude and Reactionism-Radicalism tendencies of a student group from a particular area. The number of students were 100 which is considerably large for ungrouped method of correlation. Therefore, the scatter diagram method was preferred (Table 6.4). The following are the scores (Table 6.3):

Table 6.3 : X variable (Religious attitude scores) and Y variable (Reactionism – Radicalism – scores)

S. No.	*X Score*	*Y Score*	*S. No.*	*X Score*	*Y Score*	*S. No.*	*X Score*	*Y Score*	*S. No.*	*X Score*	*Y Score*
1	85	70	26	75	59	51	73	52	76	69	48
2	90	74	27	79	58	52	74	53	77	65	45
3	76	66	28	78	57	53	74	54	78	66	47
4	81	65	29	77	56	54	73	53	79	67	48
5	87	67	30	76	55	55	72	52	80	58	49
6	88	68	31	77	57	56	71	51	81	70	45
7	92	72	32	75	59	57	70	50	82	71	46
8	65	60	33	79	56	58	75	50	83	72	49
9	70	60	34	80	57	59	76	51	84	73	48
10	85	60	35	81	55	60	77	53	85	74	47
11	76	61	36	82	56	61	78	52	86	77	47
12	87	62	37	83	58	62	79	50	87	60	44
13	78	64	38	87	57	63	75	50	88	61	43
14	80	63	39	90	60	64	80	51	89	62	42
15	81	64	40	61	53	65	55	65	90	63	40
16	82	63	41	62	54	66	56	46	91	64	41
17	84	62	42	65	50	67	57	49	92	65	42
18	63	61	43	66	51	68	58	46	93	67	43
19	89	64	44	67	53	69	59	44	94	73	43
20	70	55	45	68	52	70	60	45	95	56	37
21	80	40	46	69	52	71	63	46	96	55	35
22	85	44	47	65	52	72	65	45	97	56	36
23	73	59	48	70	50	73	66	46	98	63	37
24	74	58	49	71	51	74	67	47	99	52	32
25	71	57	50	72	53	75	68	49	100	64	54

The range of *X* variable is in between the highest score 94 and the lowest score 50 (94 – 50 = 44). This is grouped into 9 class intervals of five score units. These scores are plotted on the *X* axis.

The range of *Y* variable is in between the highest score 74 and the lowest score 32 (74 – 32 = 42). This range also is grouped with 9 class intervals of five score units and plotted on the *Y* axis (Table 6.4).

Table 6.4 : Scatter Diagram of the Scores in two tests (*X* Variable)

X \ Y (*Y* Variable)	50-54	55-59	60-64	65-69	70-74	75-79	80-84	85-89	90-94	*fy*
70-74								1	2	3
65-69		1				1	1	2		5
60-64			1	1	1	2	4	3	1	13
55-59					4	8	4	1		17
50-54			2	6	10	6	1			25
45-49		3	2	9	5	1				20
40-44		1	5	2	1		1	1		11
35-39		3	1							4
30-34	1		1							2
fx	1	8	12	18	21	18	11	8	3	100

The scores of both the tests are plotted on a dimensional sheet and tallied both test scores. The number of tallies in each cell are summed up and entered on the first column on both sides (*fy* and *fx*). The next column on both axis is an arbitrary origin (*fy* and *fx*) is marked near the centre of the distribution for working out the deviation values in the next column. The numbers in the column are +1, +2, +3, +4, +5 and –1, –2 and –3 marked above and below and a 0 near the centre. In this distribution since the origin 0 is near the centre there are 5 positive values and 3 negative values. The arbitrary origin can be fixed anywhere in the distribution on the column *fy*. Similarly, for the *x* distribution, at the bottom of the two dimensional sheet there are 5 positive +1, +2, +3, +4 and +5 and three negative values –1, –2 and –3, marked on the right and left respectively with a '0' near the centre.

Then the frequencies are multiplied by the deviation values in each interval and recorded in the columns fy' and fx'. To obtain the fy'^2 and fx'^2 values the fy' and fx' values are multiplied by the respective deviation values again. The multiplied entries on both sides were summed up and we obtain fy', fy'^2, fx' and fx'^2 values.

Table 6.5: *X* Variable: religious attitude scores

Y Variable: Reactionism-Radicalims Scores

y \ *x*	50-54	55-59	60-64	65-69	70-74	75-79	80-84	85-89	90-94	*fy*	*y'*	*fy'*	*f*(*y'*)²	*fx'y'*
70-74								20 1 20	25 2 50	3	+5	15	75	70
65-69		−8 1 −8				8 1 8	12 1 12	16 2 32		5	+4	20	80	44
60-64		−6 1 −6	−3 1 −3	1	3 1 3	6 2 12	9 4 36	12 3 36	15 1 15	13	+3	39	117	99
55-59					2 4 8	4 8 32	6 4 24	8 1 8		17	+2	34	68	72
50-54			−1 2 −2	6	1 10 10	2 6 12	3 1 3			25	+1	25 +133	25	23
45-49		3	2	9	5	1				20	0	0	0	0
40-44		2 1 2	1 5 5	2	−1 1 −1		−3 1 −3	−4 1 −4		11	−1	−11	−11	−1
35-39		4 3 12	2 1 2							4	−2	−8	16	14
30-34	9 1 9		3 1 3							2	−3	−6 −25	18	12
fx	1	8	12	18	21	18	11	8	3	100		108 Σ*fy'*	410 Σ(*y'*)²	333 Σ*fx'y'*
x'	−3	−2	−1	0	+1	+2	+3	+4	+5	137				
fx'	−3	−18	−12		21	36	33	32	15	−33 / 104				
f(*x'*)²	9	36	12	0	21	72	99	128	75	407 Σ*f*(*x'*)²				

$$\frac{\frac{333}{100}-\left(\frac{104}{100}\right)\left(\frac{108}{100}\right)}{\sqrt{\frac{407}{100}-\left(\frac{104}{100}\right)^2\times\frac{400}{100}-\left(\frac{108}{100}\right)^2}}=\frac{2.21}{2.96}=0.75$$

The next step is to obtain the value of $fx'y'$. This is the sum product of the two variables expressed by the deviation values. This is obtained in the following manner. The deviation values of X axis and Y axis in the columns and rows are multiplied as shown in the dimensional sheet and the products are entered on the upper left corner of the cells which are having the entries of the frequencies. For example, the deviation values for the X axis interval 90 – 94 is 5 and the Y axis interval 70 – 74 is also 5. These two multiplied 5 (Y') and 5(X') on the upper left of the cell ($5 \times 5 = 25$). This again is multiplied by the frequency 2 in the centre and recorded on the right bottom of each cell ($25 \times 2 = 50$). All the products recorded on the right bottom of each cell are summed up for each row. There may be also negative product in some cells. The negative values are subtracted from the summed up positive values and the balance recorded in column $fx'y'$.

Then the next step is to sum up the products in columns fy', $f(y')^2$ and $fx'y'$ and in rows fx' and $fx'y^2$. These sums are $\Sigma fy'$, $\Sigma f(y')^2$ and $\Sigma fx'y'$ and $\Sigma fx'$, $\Sigma f(x')^2$. These are the five measures required apart from the N to compute the coefficient of correlation according to the scatter diagram methods. The formula is:

$$r = \frac{\dfrac{\Sigma fx'y'}{N} - \left(\dfrac{\Sigma fx'}{N}\right)\left(\dfrac{\Sigma fy'}{N}\right)}{\sqrt{\left[\dfrac{\Sigma fx'^2}{N} - \left(\dfrac{\Sigma fx'}{N}\right)^2\right]\left[\dfrac{\Sigma fy'^2}{N} - \left(\dfrac{\Sigma fy'}{N}\right)^2\right]}}$$

Applying this formula to the data of the illustrative example:

$$r = \frac{\dfrac{333}{100} - \dfrac{104}{100}\dfrac{108}{100}}{\sqrt{\dfrac{407}{100}\left(\dfrac{104}{100}\right)^2 \dfrac{410}{100} - \left(\dfrac{108}{100}\right)^2}}$$

$$= \frac{3.33 - (1.04)(1.08)}{\sqrt{4.07(1.08)(4.10 - 1.17)}}$$

$$= \frac{3.33 - 1.12}{\sqrt{2.99 \times 2.93}} = \frac{2.21}{\sqrt{8.76}} = \frac{2.21}{\sqrt{2.96}} = 0.75$$

Having determined the coefficient of correlation it is also important to estimate the standard error (SE) of r for testing the hypothesis $P = 0$ with correlation of 0.75 and with a sample size of 100 pairs of scores the SE is computed by applying the formula $SE_r = \frac{1}{\sqrt{N-1}}$. This is to be stated alongwith correlation coefficient as $r = \frac{0.75}{\pm 0.10}$. The correlation coefficient may deviate for a sample of this size 0.10 on either side (Table 6.5).

The other estimate is the t ratio for testing the significance of a coefficient of correlation. The formula for estimating the t for r is:

$$t = \frac{r\sqrt{N-2}}{\sqrt{1-r^2}} = \frac{0.75\sqrt{100-2}}{\sqrt{1-0.75^2}} = \frac{0.75 \times 9.899}{\sqrt{1-0.56}} = \frac{7.42}{0.44}$$

$$= \frac{7.42}{0.66} = 11.24 \text{ Significant at 0.001 level.}$$

For a two-tail test with a 98(100 – 2) degrees of freedom the t value is 11.24 which is larger than the table figure 3.373. Therefore, t value 11.24 is significant at 0.001 level.

It is a tedious job to find out the coefficient of correlation by this method. Therefore, great care must be taken in calculating for accurate result. In the case of the example, the coefficient of correlation is 0.75. This shows that there is relationship between religious attitudes and reactionism-radicalism tendencies. It is observed that if the scores on RR (Reactionism-Radicalism) scale are increased there is corresponding increase of scores on the Religious attitude scale also. Those who score low on the RR scale also score low on the Religious attitude scale. These people are reactionists and conservatives and they do not want to make any change in the existing society and they are highly religious. Those who score high the RR scale also score high on Religious attitude scale. These people are liberals and radicals and they are less religious or anti-religious in their attitudes. The scores in between these two (high and low) levels of scores indicate the neutral position. These people are moderate in both the measures.

Spearman's Rank Order Correlation Method

The Pearson's product-moment correlation method described above is preferred by us whenever we find it suitable to our data. It will not be always possible to apply Pearson's Product-moment correlation method for all the data. There are certain data to which Pearson's correlation method does not suit. In such cases some other method of correlation is applied to. The Spearman's rank difference correlation is one of the methods very often applied especially for qualitative variables and when samples are small. This method is very convenient especially when the number of cases are less than 30. This method disregards the values of the measurements, but takes into consideration only the ranks of the cases. In every day life we may come across many cases of ascertaining the places of persons and things. If comparison between the performances of two persons is made one finds the first place and the other the second place. Similarly, when we compare different objects we place them in the order of preference on the basis of their qualities.

Further, wherever there are more number of persons working a rank order with regard to their abilities and efficiencies takes place naturally. Employers make a rank order of the employees in workshop, industry and office. Students in the classrooms are rank ordered by the teacher on the basis of their performances. Sportsmen, tournament players are given rank orders, cinema stars are given order of preference on the basis of their different roles they play and on the basis of their talent. For all such cases, Spearman's rank-order correlation method is very much useful. In this method of correlation, the data is presented in sets of ordinal numbers such as 1st, 2nd, 3rd etc. This again is replaced by ordinal numbers as 1, 2, 3, etc.

The data for correlation are in terms of scores. The first step, therefore, is to convert them into rank orders. A computation of the correlation coefficient is given here as an illustrative example. In a classroom there were 15 students and two tests were administered on them. One was the aptitude test and other was achievement test. The purpose of this test was to see the level of association between students' aptitude and their achievement. The aptitude test items were answered with 'Yes' and 'No' answers and the achievement test items with 'Yes', 'No' and '?'. The tests were scored according to the procedure and each individual's scores on the two tests were elicited.

The individual's two test scores are given in two columns. Then the ranking was made. This was made from the highest to the lowest for the first test scores. The highest score received rank 1, the next highest 2 and the third highest 3 and so on. But there is some difficulties if two or more individuals received the same score. For example the first and seventh persons received the same score 45 and similarly the sixth and fifteenth persons received the same score 39 in the first test. In such case we must assume that one of them should come first. But we did not know who he was. Therefore, it is better that we should take the average of the two ranks 11th and 12th as 11.5 for both individuals. The next rank is 13. But there is tie again. The sixth and fifteenth persons received the same score 39 in the first test. Therefore, the average of 13th and 14th ranks should be taken for both as 13.5. If there is any tie in the individual scores there may not be any regular ranks. Giving the average among the ties the scores fifteen persons were ranked from 1 to 15 for the first test.

Table 6.6 : Rank difference correlation coefficient between the scores of aptitude test and achievement test for 15 students

S.No.	*Aptitude Score*	*Achievement Score*	R_1	R_2	*D*	D^2
1	45	73	11.5	7	4.5	20.25
2	69	77	4	6	2.0	4.00
3	50	83	9	5	4.0	16.00
4	46	48	10	14	4.0	16.00
5	33	47	15	15	0.0	0.00
6	39	57	13.5	12	1.5	2.25
7	45	68	11.5	9	2.5	6.25
8	74	89	2	3	1.0	1.00
9	70	91	3	1	2.0	4.00
10	63	85	6	4	2.0	4.00
11	57	68	7	9	2.0	4.00
12	76	90	1	2	1.0	1.00
13	66	52	5	13	8.0	64.00
14	53	68	8	9	1.0	1.00
15	39	66	13.5	11	2.5	6.25
					D^2	150.00

The scores of the second test also were ranked in the same way as in the case of the first test. In the second test three individuals, the 7th, the 10th and the 14th persons received the same score 68. They occupy 8th, 9th and 10th places respectively. Taking the average of these three positions $(8 + 9 + 10 \div 3 = 9)$ we assign the rank 9 to all the three persons leaving out the ranks 8 and 10 in the list. The next step was to find out the difference (*D*) in the ranks for each pair by subtracting the smaller rank from the large irrespective of the place. Sign is not taken into consideration here. Then the rank difference is square up (D^2). The squared rank differences are summed up (ΣD^2). Then the Spearman's rank difference coefficient of correlation formula is applied to the data:

$$\gamma p = 1 - \frac{6\Sigma D^2}{N(N^2 - 1)}$$

Where the ΣD^2 = the sum of squared differences between the ranks. N = The number of cases. In the case of the illustrative example:

$$\gamma p = 1 - \frac{6 \times 150}{15 \times 224} = \frac{900}{3360} = 0.267 = .27$$

The Spearman's rho coefficient is almost close to the Pearson's r, but a little lower than the Pearson's r. There is no formula that can be generally applied for determining the confidence limit of the Spearman's rho. We cannot apply the standard error formula $\frac{1}{\sqrt{n - 1}}$ to our example as the sample is less than 30. But from the table (Table 6.6) we can determine that the coefficient .27 is significant at 0.01 level.

PART II

INFERENTIAL STATISTICS AND MEASUREMENT

INTRODUCTION

Inferential statistics are the statistical processes by which conclusions are drawn in some way from certain facts or premises. By applying the inferential statistics one can obtain conclusions in his study. The researcher makes conclusions by reasoning from something known or assumed. The logic of induction and deduction are applied in inferential statistics. It is also one way of prediction. In most of the studies conclusions are not drawn directly. Conclusions are obtained indirectly. In applying the inferential statistics in social and behavioural sciences the probability of occurrence of events are predicted.

In studying some aspect of the behaviour of a population, it will not be possible to study the whole population. Only a portion of the population is chosen, and the results are attributed to the whole population. But what portion of the population or what percentage of population is to choose for the study? This is the problem that every researcher faces in conducting an investigation. The sampling methods try to solve this problem. In studying a problem, hypothesis is formulated and tested and the errors involved in this are also determined.

In testing a hypothesis the inferential statistics provide several methods, like the chi square, the *'t'* test, the critical ratio, Duncan's Range Test, and the analysis of variance. The researcher can make use of any one of them and test his hypothesis. Some of the methods can be used in small group studies and some can be applied in large and multigroup studies. Sometimes, the researcher may have to compare the behavioural aspects of several groups. The analysis of variance and Duncan's Range test may be applied in such cases. The inferential statistics is also applied in estimating the errors involved in measurements.

7

INFERENTIAL STATISTICS

Sampling Methods

The subject-matters stated in the preceding pages are based upon descriptive statistics. They have been dealing with thc samples of measures precisely and accurately. But the subject-matter of psychology cannot be fully dealt with descriptive statistics only. Certain matters go beyond the descriptive statistics. The researcher wants to make predictions in terms of general principles. Further it becomes also important for him to know the nature of the interpretation of his data. Particularly, the researcher is interested in the confidence level of the conclusions of his work and from which he wanted to generalize the traits or characteristics of his sample and ultimately the population. The statistical techniques that can facilitate the process of generalization are called inferential statistics. The basic subject-matter for all inferential statistics is the sampling, mostly representative sampling.

Sampling

The researcher when he plans to include people in his investigation he will have to decide first of all whether to study the whole population of people or to study only a small group of people taken from the large population. A small portion or group of people taken from the large population is called a sample. The process of selecting those people from the large population is called sampling. The large population in other words may be called the universe. According to D'Amato (1976) "a population is any complete collection of objects or things; a sample is any subtotal portion of a population." There are several types of sampling. A sampling plan should specify how the elements are drawn from the large parent population and also how many elements are drawn. Blalock (1960) states that all sampling plans can be brought under two broad categories such as (i) non-probability sampling and (ii) probability sampling.

The Non-probability Sampling Plans

These sampling plans do not provide any basis for estimating how far the sample characteristics identify with the parameters of the population from which the sample was drawn. Actually the researcher will not be in a position to identify the larger or parent population. Therefore, samples drawn according to non-probability sampling plan lacks the characteristic of generalization. Since the non-probability sampling plans are not designed for the functions of inference there is no use of discussing sampling errors involved in such samples. But there are certain special situation in which the researcher may use the non-probability sampling plan alongwith probability sampling. Some investigators justify the non-probability sampling plan by including some elements randomly in their investigations and state that by which the characteristics of generalization is improved. But to what extent this procedure qualifies the non-probability plan to be called as probability plan? However, those who are interested in generalization or inference should know the limitations of the non-probability plans and be cautious since there are several non-probability sampling plans. According to Coleman (1958-59) the non-probability sampling plans include, (1) systematic sampling, (2) quota sampling, (3) accidental sampling, (4) purposive sampling, (5) snowball sampling, (6) saturation sampling, (7) dense sampling. Selltiz (1967) and others discuss these sampling plans.

It is also true that many of the researchers in the field of psychology often resort to the non-probability sampling plans and generalize their results not knowing the limitations of this plan. A brief discussion of these plans may enlighten the investigators who are interested only in obtaining a certain number of elements to satisfy their limited research aims.

The Systematic Sampling Plan

This sampling method has certain characteristics of randomness and also has some traits of non-probability. It is the procedure of obtaining the elements by drawing every *n*th the person (every fifth person, every tenth person) from an already determined list of persons. For example, drawing every fifth person or tenth person from a city telephone directory is a systematic sampling. Though this procedure appears to be like probability sampling, the results of this method is sometimes misleading the investigator. Though this method has certain advantages that the sampling is easy and quick in the collection of data, it has certain disadvantages. This method ignores the chances of all persons between every 5th or 10th person chosen. According to this method only the first person is selected at random and also the other successive 5th or 10th persons are fixed. Persons in between 5th and in between 10th do not find chance to be included in the sample. This is how the non-probability systematic sampling lacks the characteristics of probability sampling characteristics.

Another disadvantage of this plan is that it tends to increase the potential sampling error. The list is prepared according to some order. The list may not be having every trait of the elements proportionately. For example, age, sex, education, income, etc. of the elements may cause certain amount of bias. There will be more elements from a particular age or sex and so on. Since each element is not having equal chance of being included in the sample it will not be a random sampling. Hence a systematic sampling not a random sampling.

The Quota Sampling

This sampling method like the systematic sampling mainly lacks the characteristics of randomness. It is by selecting the required number of elements of those mostly accessible to the investigation with the characteristics in which he is interested. For example, the researcher may state that he wants to collect data from certain number of elements of a particular religious group or from a particular area. Since certain specific elements are there in the sample, Black and Champion (1976) call this as poor man's proportionate stratified sample. But it cannot be called as proportionate stratified random sampling plan.

This sampling plan also has certain advantages and disadvantages. The main advantage is that this method is less costly than any other sampling methods and quick and also satisfies the objective of the researcher.

One of the major disadvantages of quota sampling method is that the researcher takes too much leniency in selecting the most accessible elements, including his friends and relatives. Further, there is no means of establishing randomness in this method. The resulting sample is not of a probability sample. Therefore, the results are not fit for generalization to any extent since the population is not known in this method.

Accidental Sampling

The researcher in accidental sampling method is not interested in including the elements with any specific characteristics. He accepts whatever, he can get and satisfied with it. This is a crude sampling method and no generalization can be made from the sample obtained by this method. The researcher uses this method to obtain the opinions of the questionnaire he is developing for some investigation. Therefore, he gets the opinions of his accidental respondents to express their views or comments on the items, words, size of the questionnaire, clarity, time and emotional reaction and so on.

This method is mainly used by media reporters like, the radio, television, newspapers who are moving about from place to place for information. The reporters of these media simply ask the people on the streets, in the shopping centres, in working places to express their views on certain issues or about the information to be revealed in their media. This is some what similar to quota sampling method. But it lacks the characteristics which the quota sampling method includes in the sampling.

The advantage of this method is that it is more convenient for collecting samples and also more economical and time saving. The disadvantage of this method is mainly that it is more biased as it is more subjective and based upon the researcher's interest and convenient. There is no chance for probability sampling in this method. Some call this as incidental sampling method.

Purposive Sampling

In purposive sampling method the researcher includes specifically certain element and collects only those samples which possess the particular trait. Since this type of sampling is very much selective all the relevant information are represented in the sample. The researcher makes his decision with regard to the typical population from which he presumed to collect the required information. Because of this, the method is also called judgment sampling method. He is of the opinion that by carefully selecting certain elements and obtaining information from them the result may have the same characteristics as the result obtained by some probability sampling method.

The advantage of purposive or judgment sampling method is that it is less costly and easily approachable. This is somewhat similar to quota sampling method, but much better than that, with regard to the care paid in selecting the samples that the researcher seeks for his objectives. Purposive sampling assures of the elements that are important for research.

One of the serious disadvantages of this method is that the researcher takes risk in generalizing the results. Because there is no reason to state that the sampling is random or representative. In purposive sampling the presence of the specific elements is more emphasised. Another important disadvantage is that inferential statistics cannot be used in this non-probability sampling plan. The fundamental principle of inferential statistics is that there should be randomness in the sampling technique. Purposive sampling as a non-probability sampling plan lacks the randomness.

Snowball Sampling

The snowball sampling method is mainly used by researcher who are interested in small group research. This is a way of obtaining information among individuals who are having informal social relations. The researcher includes all individuals in a limited or small group and identify their friends and associates till he (the researcher) achieve a social pattern. According to Coleman and others (1957) a plan of spreading out of information is revealed through snowball sampling. *For example*, how the information about a new effective medicine were spread out among the physicians? Was it through a medical conference or by having contacts with other physicians informally, or through medical bulletin? An important information was widely spread out among the members of a given group, the physicians. Answers to these questions may be obtained through snowball sampling method. It is mainly a sociometric pattern.

The snowball sampling method has certain advantages over other methods in small group research such as businessmen, industrial workers, physicians, engineers and so on. The respondents of snowball sampling method are subjectively chosen and the researcher cannot use the probability or inferential statistical method in this. When the number exceeds certain limit it is very difficult to apply this method.

Saturation Sampling Method

The concept saturation means thoroughly soaked. The sampling according to this method includes all the elements of a given population with the characteristics in which the researcher is interested. For example, including all the engineers of a small community or all the school teachers of a small group or all the telephone operators of a small group in the sampling may be called saturation sampling method.

This method has certain advantage over other methods. It is possible to state what the particular group as a whole views the problem. It straightaway reveals the fact and not depending upon inference.

Dense Sampling Method

In this method, the concept dense is used by Coleman (1958-59) in the meaning of crowded together or 'densely'. The researcher includes more than one-half or majority of respondents in the sampling with the specific characteristics in which he is interested. For example, out of 1000 lawyers the sample may be consisting of 500 to 600 lawyers. This size of sample is much greater than any of the probability sampling plans. Generally probability sampling plans require ten per cent or 20 per cent of the population which comes to 100 to 200 lawyers. But the dense sampling method cannot be used if the population size is very big. This method is more successful in small populations.

These non-probability sampling methods discussed above are not invalidated thoroughly in psychological researchers. The only main shortcoming is that they lack randomness and any application of inferential statistics is to be with risk and predictions may not be accurate. But these methods are in usage in psychological researches and researchers should bear in mind that they should apply only descriptive statistics to the data obtained by these methods.

Probability Sampling Plans

Many of the statistical tests and inferences mainly depend upon the data collected by the probability sampling methods. According to probability sampling methods there is a probability or chance or likelihood of inclusion of every element/person of the population in the investigation. The researcher who applies the probability sampling technique should bear in mind that the *size of the parent population* or universe from which the sample is obtained should be known to him. Secondly, he should state definitely the *size of the sample* he is going to draw from the population. This, he should mention in number and also in the percentages. Thirdly, he should see that *each element/person* in the population should *have an equal opportunity* or chance to be included in the next sample.

But these conditions of probability sampling plans cause a big problem to a single investigator, particularly to know the size of the parent population and to provide equal chance to every element in the next sample. If the concerned population is small, say 200 or little more he can satisfy the conditions. He can provide chance for all 200 persons to be included in the subsequent sample. Each individual has 1/200th of a chance to be included in the sample. In the same way, if the population, for example, is 20,000 each individual in the population has 1/20000th of a chance of being included in the sample. The practical problem is that it is absolutely impossible for any investigator to state the size of the population correctly as many people die everyday and many are born every day and many migrate to foreign countries and many immigrate into the country. Further, whatever, the population size is, each of them must be included in the subsequent sample. Therefore, the identification of the population size and the inclusion of each element in the subsequent sampling are quite impossible for a single investigator who desires to fulfil his objectives of his research. In this context a remedy to the problem has been suggested by the statisticians. A working probability sampling plan is to have representative samples and through which a generalization is drawn and the results of the sample are also attributed to the parent population from which the samples are drawn.

The investigator can state how far his representative sample may closely resemble the characteristics of the parent population and also how far the characteristics of the parent population are revealed. The statistics applied to assess this is called sampling statistics. Statisticians call this method as *inferential statistical method.* This method permits the researcher to infer the characteristics or traits of the parent population by analysing the characteristics of the sample drawn from the parent population.

Statisticians believe that by applying the probability sampling plans it is possible to estimate error accurately that occurs in the estimation of the characteristics of the parent population by analysing the characteristics of the sample drawn from the parent population. There are several probability sampling plans. It is important to discuss them here. They are: (1) Simple random sampling method, (2) Proportionate stratified or controlled random sampling method, (3) Disproportionate stratified random sampling method, and (4) Cluster or area sampling method.

What is Randomness and Random Sampling?

According to Peatman (1947) the important factor that regulates all probability sampling plans is the randomness. The meaning involved in randomness is that all the elements/persons in a population should have an equal chance of being included in the sample.

The random sampling procedure provides a guarantee that all the elements/persons in a population has an equal chance independently and proportionately of being included in the samples. The procedure is by selection. The selection is like tossing a coin or a disc. A simple and most familiar type of sample selection is by putting each number in a separate slip and collecting them in a large bowl. All the numbers should be included in this. Each number stands for an element or person in the population and the whole population members have been numbered. Thus, every member of the population is representing in the slips collected in the bowl. Then the slips should be thoroughly mixed and the researcher should select one slip at a time blindfolded. He should continue this till he is satisfied with the size he wanted. In this way the random sampling selection is made. Whatever, type of criticisms levelled against this method, this is simple and easy method. But there are also other methods of random selection.

A table of random numbers, published by the Rand Corporation (1955) is recommended by many experts for sample selection. This table of random numbers is a table of numbers in rows and columns.

There is no order in these numbers and a number does not appear often in this table (*see Appendix A*). All the numbers are grouped in twos of five in a row and twos of five in a column. There are 25 pairs of numbers in each square. The table has no other special significance except that the numbers are conveniently arranged without frequent repetition. While using this table, the researcher should state how many elements/persons are there in the population of his study. Then he should enter them one by one from 1 to *N*. The *N* is the total number of elements/persons in his population study. Suppose, for example the population has 900 elements. The researcher, then should number all elements from 1 to 900 and he should decide the size of his sample he would like to have, whether it is 10 per cent or 15 per cent. Let us assume that he decides 10 per cent which means 90 elements from the population. Hence, the sample size is 90. Next he has to enter into the random numbers table and without looking at the numbers, has to place his finger on any number anywhere on the page. The researcher will have to begin from the place where he puts his finger. Since the population size is 900 and it contains three digits. If the population size is 5000 then it is four digits and if 20,000 it is five digits. Therefore, for the example he should take into consideration groups of three digits in the table of random numbers.

Generally everybody may start on the upper left hand part of the table. But he can start from any place where he puts his finger as stated earlier. The first three digits number in the table is 100. This indicates that the person numbered 100 in the population should be included as an element of his random sample drawn. The researcher must make an entry in his record book. He further moves *down the column* and finds a three digits number 375 and the person numbered 375 is included in the sample. While moving down the column for the next sample, the three digits number the researcher finds is 084. The first zero is ignored and the number 84 is taken into account and, therefore, he includes the person numbered 84. Again, for the next sample he moves down the column and finds a three digits number 990. There is no such numbered element in the population as the population size is only 900. Therefore, the zero is ignored and the number is 99 and the person numbered 99 is included in the sample. The zeros in the front of the number should be ignored. If the zero is at the end and including the zero, if the number is greater than the population, then the zero should be ignored and the remaining number should be included in the sample. The same procedure should be followed till the investigator gets sample size 90. If the first three digits numbers of the table are over, that is on reaching the bottom of the column, he should move to the top of the column and move one digit to the right, omitting the first number and starts the next three digits numbers. The next three digits number in the table is 009. This refers to the person numbered 9 (zeroes should be ignored). The next four numbers 754, 842, 901 and 280. The persons numbered 754, 842 and 280 are included in the sample, but there is no person numbered 901 since the population size is 900. Hence it is ignored and the researcher should keep on moving systematically till he gets the random sample size of 90 elements.

The researcher, while collecting the samples of numbered persons can skip over 2 or 3 numbers each time and also change the columns on the right hand part of the table and he should not move either sideways or diagonally. On any account he should not deviate from the systematic procedure.

One of the advantages of the random number table is that it is a perfect model for obtaining random samples from small populations. The other advantage is that it is easily administered. The disadvantage of this procedure is that it is very laborious and cumbersome if the population size is more than 20,000 elements. In such cases it is better to make use of computerized random sampling method.

Simple Random Sampling Method

All the characteristics described above are applicable to the simple random sampling method. Mostly in all sampling plans with small populations this sampling method is included by the investigators in all social sciences. This method ascertains that every element/person of the population has an equal and independent chance to be included in the sample and in the subsequent samples. Equal chance means that an element in a population of 1000 has a chance of being included in the sample 1/1000th of a chance. In a population of 2000 every person has 1/2000th of a chance of being included in the sample. This is according to the statistical principle. But due to certain reason if some persons are not purposely included in the sample by the investigator then the obtained sample is not a random sample. Researchers sometimes exclude certain elements from their investigation as they face some practical difficulties in collecting the sample according to the principle.

For example, a sampling from some socio-economic status groups, sampling in the areas where there are social, economic and emotional deprivations, in certain geographical areas where there is instability of residential facilities, in areas where communal, racial and religious prejudices are rampant is very difficult and the investigator is facing risk in taking samples. As a result of this the investigator will have to skip over certain elements in his samples. In a mixed population it would not be possible to identify certain elements with the original population. In such cases even though they are excluded from the sample it will not affect the sample results.

It has been stated above that every element has an independent chance of being included in the sample. *An independent chance means that the draw of one element from the bowl will not affect the chance of another element of being included in the draw* we are making subsequently. For example, in a population of 900 the fishbowl draw is used to collect the random samples. After each draw few elements are remaining in the bowl and the chances for them are greater if we do not replace the drawn elements in the bowl. After the first draw from the 900 elements, one is removed from the bowl, leaving 899 elements in the bowl. First each element has 1/900th of a chance of being included in the sample. After the first draw the remaining elements will have 1/899th of a chance of being included and after the second draw the remaining elements will have 1/898th chance and after a third draw the remaining elements will have 1/897th of a chance and so on. Since we are not replacing the drawn elements in the bowl again the procedure is called *random sampling without replacement.*

But, we can also replace the drawn element once again in the bowl before a second draw is made and the drawn element may have *another chance of being drawn once again* and included in the sample. This procedure is called *random sampling with replacement.* This chance depends upon the size of the population. If the population is considerably large a second chance for the drawn element is very remote. Even if it is drawn for a second time we can ignore it and need not include in the sample and place it again in the bowl. By chance if comes once again it can again ignored and need not include in the sample and can be placed in the bowl once again.

The simple random sampling method also has certain advantages and disadvantages. It provides chance for every element in the population to be included in the sample. Therefore, the sample becomes a representative sample. Secondly, the simple random sampling method is the basis upon which all other types of random sampling methods are designed. Thirdly, the researcher need not know the correct population size before planning the sampling. Fourthly, the sampling can be easily estimated from the samples drawn from the population. Fifthly, since the researcher need not know the characteristics of the population before the collection of sample he does not make the classification of errors. He neglects the classification error as he uses the simple random sampling method.

This method fails to make use of the investigator's knowledge about the population. Secondly, this method does not include certain elements of the population in the sample. There is chance for greater sampling error in simple random sampling method compared to the stratified random sampling method of the same sample size. It is because of the presence of hetrogeneous characteristics in the sample the error chance is greater.

But the researcher can reduce the sampling error in this method by knowing more about the characteristics of the given population and also by the use of his knowledge in drawing more representative samples from the population.

Proportionate Stratified Random Sampling Method

The researcher is supposed to know many of the characteristics of the population he studies. For example, he may be knowing the characteristics such as age, sex, qualification, the area, cultural variables such as race, religion and community. Therefore, he may choose a proportionate stratified random sampling method. In this method the sample may have the specified characteristics in correct proportion in the same way the characteristic spread out in the population. This method ensures better representation of the samples. It is a kind of experimental control.

Out of several characteristics of the population the experimenter chooses any one or more than one characteristics and assigns for his sampling. His collection of sampling pertains to one single characteristic or more than one and the sample is supposed to possess the specified characteristic or characteristics in which he is interested. It may be also called as a sub-group in the population identifying the same characteristics as the parent population possesses.

Suppose, for example, the researcher wants to collect a sample of 20 per cent of students from a college of various classes to study their attitudes toward higher education. It is a degree college having only three year courses. The college authorities admit in the first year 500 students and due to drop out, in the second year the strength of the students was 300 and in the 3rd year the strength was 200. On the whole he has to collect a sample of 200. This is 20 per cent of the total strength and also 20 per cent of each class strength. Since the same percentage from each class is maintained the sample is proportionately distributed. The composition of the college population is 500 + 300 + 200 = 1000. The proportion of students in each class is worked out.

Table 7.1 : Showing the composition of population and the distribution of proportion students in every year

Year	*Population*	*Proportion*	*Sample Size*
First Year	500	0.50	100
Second Year	300	0.30	60
Third Year	200	0.20	40
Total	1000	1.00	200

The proportion is obtained by deciding the class population by the total population 500/1000 = 0.50, 300/1000 = 0.30, 200/1000 = 0.20. Hence the sample consisted of 20 per cent of the total population proportionately distributed in all 3 classes. The researcher includes 100 first year students, 60 second year students and 40 third year students in this sample. This process is called stratification. The researcher, now can apply the *simple random sampling method* for drawing the required number of elements from each class or sub-population. In the first year class there are 500 students and he needs 100 students from this class for his sample. He first numbers all 500 students in slips from 1 to 500 puts them in a bowl. Then by using the table of random numbers the researcher selects the sample. Since the first population is 500 he has to use three digits numbers in the selection. He can follow the same procedure described earlier. He can start anywhere in the page of the table by placing his finger blindfolded. However, we shall see one example for illustration. Let us start from the upper left hand corner. Since we need three digits, the first number we see in the table is 100. The element numbered 100 is to be included in the sample. The next three digits number is 375. The element numbered 375 out of 500 is included in the sample. This procedure he can continue till he gets the required sample 100 from the first group. He can follow the same procedure for other sub-groups also. After completing all three sub-groups he can combine each class simple random samples into a proportionate stratified random sample of college students. If more independent variable are there the proportionate stratified random sampling becomes complex.

In spite of all these, the proportionate stratified random sampling also has certain disadvantages apart from advantages. This method assures of the representativeness of the parent population. The sample drawn according to this method will be a better estimate of the population characteristics. In this respect it is better than the simple random sampling method. The sampling error is very much reduced in this method compared to the simple random sampling method. It is not necessary to weight the elements according to the distribution of the population in this method. For example, in the case of the size of male and female elements in the population and the difference in their age need not be taken into consideration in this method.

The main disadvantage is that it is difficult to obtain samples by this method if the population is very large and categories are many. It takes more time to draw elements from each of several categories. The researcher is expected to know in advance about the composition and distribution of the population characteristics. There is a possibility of classification errors in this method if there are several strata to be identified. While categorizing elements in the population the elements are sometimes put in the wrong places and this leads to classification errors and may lead one to a wrong interpretation of the results in the final stage. Therefore, the researcher must be extremely careful in assigning the elements to different categories and also in stratifying the characteristics. If the stratification is made inappropriately then there is a commitment of bias in the sample which may increase the sampling error associated with it. Comparing with simple random sampling method the stratified random sampling method is more costly and time consuming.

Disproportionate Stratified Random Sampling Method

This method is quite similar to proportionate stratified random sampling method in some respects. There may be main strata and substrata. Disproportionate stratified random sampling does not pay more attention on the proportionate distribution of the population. Some sub-strata may have more representation and some may have less representation.

Suppose, for example, age may be used as a strata and the researcher distributes the population of 200 persons into five sub-strata on the basis of different age levels such as 20-29, 30-39, 40-49, 50-59, 60-69. The following is the break-up percentage of the distribution of elements among the five sub-strata.

Table 7.2 : Showing the disproportionate stratified distribution of the population on the basis of sub-strata

S.No.	*Sub-strata*	*Percentage*	*N in each Strata*
1	20-29	30%	60
2	30-39	25%	50
3	40-49	15%	30
4	50-59	15%	30
5	60-69	15%	30
		100%	200

If the researcher wants to make this a proportionate stratified random sample he should make the sample consists of five equal parts, each part having 1/5th from each of the strata. That is, if he wants a sample size of 200 then he must draw 40 elements from each strata. One-fifth of the sample is 20 per cent. There is no strata with 20 per cent of the sample. The first-two are over represented and the last three are under represented. By this means the researcher has not given equal representation to all the five sub-strata. Therefore, he gives greater weight to certain elements that are not adequately represented in the population. This is because of the disproportionate distribution in the population.

This method has certain advantage with regard to the timing. It takes less amount of time compared to the proportionate stratified random sampling method. The researcher need not be so careful in drawing specified number of elements in every sub-strata of the sample as in the case of proportionate stratified random sampling method. The investigator can give more weight to certain elements which are not having adequate representation in the sample in comparison with other elements.

This method deprives of proper representation to each sub-strata. Some sub-strata may represent more heavily than others. This causes some amount of bias and hence there is less representation. As in the case of proportionate stratified random sampling, in this method also the researcher is required to know the composition of the population. Any wrong classification of the elements by the researcher may result in classification errors. Therefore, the researcher should have adequate knowledge about the population of his study.

Cluster or Area Sampling Method

This is a type of probability sampling method mainly used in the field of agricultural researches in determining the effects of various fertilisers and soil treatments. The various planning methods on crop yield resulted in the application of a sampling method and the random areas of crops. A selected area may be made into several divisions by a number of vertical and horizontal grid lines drawn over the total area. The various divisions are numbered and selected randomly using the random number table. When the sample selection is made randomly the researcher can determine the crop yield and compare it with crop yield for another sample of division selected from another area.

This method is very extensively used in social science researches specially in *field research, ecological studies and population studies.* Just like the grid lines make divisions in agricultural researches, geographical divisions are made on the map and the divisions are numbered and certain divisions are drawn randomly as random sample. The divisions may be in territory, community, neighbourhood and so on. Then the researcher can interview all the persons living in the specified division. It may be a city or urban or semi-urban area or typical rural area and so on. This is how the cluster sampling method is applied.

It is better to illustrate this with an example. Suppose, for example a researcher wanted to assess the attitudes of the residents in a particular state towards family planning and birth control. It is physically impossible to assess the opinions of all the residents in the state. It involves more expenditure, needs more manpower for conducting the interviews. And also the residents of the state should provide us with information that are comparable. If the area is with fully rural backgrounds the work becomes more difficult. Interviewers will have to move long distances by walk from interview to interview. Therefore, it is not encouraging the sampling plans stated in the preceding sections. Hence a different method is to be followed.

The researcher by drawing horizontal and vertical grid lines and divides the whole state into several divisions in the map. Then, he numbers each division from 1 to *N*. Here *N* refers to the total number of divisions he made by the grid lines. He selects certain specified number of divisions for his study by using the random numbers table. He conducts interviews in these divisions. All the persons living in these divisions which were randomly drawn are interviewed. There may be variations in the divisions. In certain divisions there may be few residents say 5 and in certain divisions there may be 30 residents. All the residents are grouped together or clustered in a given division and interviewed and added in the final data collection. Suppose, if the randomly drawn divisions have excessively greater number of residents then the researcher may draw another set of divisions randomly drawn from the divisions originally made. Now he subdivides every division selected randomly as was done in the beginning and draws a number of divisions randomly from the first original division. Every division drawn is subdivided and samples are selected from them. This procedure is stated as *second stage units*. A third stage units also is possible sometimes. The researcher first makes divisions and make random selection for a specified number of divisions according to the procedure. While sampling with a division he observes certain irregularities like excessively large number of elements in the divisions. In such case he once again draws another set of divisions and again subdivides them and draws samples from randomly selection subdivisions. If this procedure is continued for a second or third stage it is called *multistage sampling*.

The area or cluster sampling is very advantageous to apply on a larger populations for the collection of data. The researcher need not know the composition of the population in the area he studies. He conducts the interview on the residents of the divisions randomly drawn from the geographical divisions. The sample obtained in this manner is called probability sample. It is less costly, less time consuming and the respondents may be substituted at once. Sampling is made in a clustered manner and not individually. It is more flexible than other method by having second stage or third stage units in sampling.

The disadvantage of this method is that each sampling included in the study is not of equal in size. The researcher may not have control over this characteristic. This leads to bias in the sample and sampling error. The individual who participated in one cluster on one occasion may participate in another cluster on the next day when he moves there. The method does not prevent such repetitions.

Conclusion

From the foregoing discussions one will have to bear in mind that any research or investigation should be planned properly and select the appropriate population and study the characteristics of this population as far as possible before he decides to draw samples from that population. Then he will have to decide his sample size and the sampling method by which he can collect the correct information required to fulfil his research objectives.

The sampling methods we have discussed here may appear to be having certain similar characteristics. But each one is different from the other in some respects. Every sampling method has certain advantages and also certain disadvantages. Therefore, it is for the researcher to understand the merits and demerits of the methods and associate it with his plan whether it could be suitable for his purpose.

Probability, randomness and stratification are the important processes involved in the sampling method. The non-probability sampling plans are devoid of randomness. An element/person in a population is not having equal chance of being represented in the sample. If randomness is absent in a sample then there is a chance for bias in the selection which becomes the cause of sampling error. Instead, the probability sampling methods put more emphasis on randomness and proper representation of the element in the sample and thereby the level of sampling error is reduced considerably.

Therefore, an investigator who is interested in making his research results accurate and more reliable he must select the appropriate sampling method and follow the correct procedures and draw adequate samples.

8

PROBABILITY AND CHANCE

What is Probability?

Individuals in groups differ from one another in almost every measurement. Individuals not only differ from each other, but they also differ among themselves if measurement is made on two different occasions. For example, the weight of a person measured in the morning is different in the evening or on the next day if measured once again. We also know the reason for this type of difference in the measurement. After heavy meal one may naturally weighs more. Therefore, individuals when measured on different occasions vary considerably. Similarly statistically derived measurements on some samples of a population may also vary. Hence, difference between two or more measurements is always immanent.

The psychological measurements whether it is intelligence, or achievements, or aptitudes, or reaction time will not be the same for the individuals when one variable measured on two occasions, one measurement in the morning and the other in the evening. But these two measurements will not be the same, they vary. For example, an achievement test was administered on a sample of 100 Higher Secondary School students and the mean score was found to be 95. The same test was administered on another sample of 100 higher secondary students expecting the same score. But the mean score of the second sample was 96. Since the difference is only one point we could have said that it was a just chance difference. We might have accepted this decision. But if the difference between the means of the two samples was 5 points, could we say the same reason that the difference is 'by chance' in this case also. Then what about the occurrence of the difference of 10 points or 15 points between the means of two samples. How much difference between the means of the two samples could occur or how far the two means could differ for rejecting the hypothesis on the ground that it is by mere chance the difference has occurred?

In another study 40 college students made an expedition on mountaineering on two routes *A* and *B*. These routes have the same distance and have the same difficulties from the starting point to the terminal point. But they will have to go 50 yards on a common path and then see the choice routes *A* and *B*. The choice was left to the decision of every student whether it is route *A* or *B*. We can observe the behaviour of the 40 students at the choice point of route *A* or *B*. Our expectation is that the chance of route *A* should be equal to the chance of route *B*. If so 20 students choose route *A* and 20 students should choose route *B*. This is the probable occurrence. But there would not be any surprise that 22 students choose route *A* and 18 students choose route *B*. And also there is no wonder if 30 take route *A* and 10 take route *B*. We expected 50-50 chance for routes *A* and *B*. But the occurrence is different. Could we attribute this occurrence to chance? How far the sample could deviate from 50-50 chance so that we can conclude that the result of the behaviour of the students was not due to chance? These questions raise the statistical problems of the level of confidence we have in the means, standard deviations, standard errors of the means, correlation coefficient, errors of measurement and other statistical derivations from the samples. Statisticians have provided with a method called test of significance to deal with such problems.

The test of significance could enable us to estimate whether the observed differences in the samples are due to chance or something else. This statistics enables us to decide whether the hypothesis can be accepted or rejected. We have discussed these aspects elsewhere in this treatise. But as the fundamental processes for all those aspects, we must know something about the characteristics of probability and chance.

Probability, first of all, is a statistical concept. It refers to a measure of the relative frequency or likelihood of occurrence of an event. Ordinarily values lie between impossibility or zero and certainty or one derived from a theoretical distribution or from observations.

Probability refers to the condition of being probable, an event that is probable. Probable means that something is likely to happen, but not necessarily and also most likely to happen. For example, a person is probably to be chosen for a post or for a team to take part in the events. But it is not sure that it will take place. It may happen or may not happen, both are probable. The chances of occurrence are between zero and one.

Morris Cohen (1934) says that there are two ways of looking at probability. The first view is that probability refers to the statement that what is stated is true. The probable truth behind this should be evaluated or tested by a sample. The second one is fundamental to sampling process, is essentially mathematical. According to this view, probability expresses frequency of the occurrence of a given event and it is relative to the frequency of non-occurrence of that event. Therefore, in any series of events there can be occurrence and non-occurrence. In other words there can be true or false. Probability is theoretically expected frequency of occurrence of any event when the event is repeated for a number of times by some procedure. For example, flipping a coin for a number of times may result the falling of head and tail. We do not know whether there going to be head or tail for a flipping. It is done without any motive and there is no bias. Therefore, we may expect the occurrence of head as many times as the occurrence of tail. The probability of frequency of occurrence of heads should be: $P = \frac{h}{N} = \frac{t}{N}$. Here the N is the number of times the coin is flipped; the h, the number of times the head occurred; and t, number of times the tail occurred. If the coin is flipped 20 times then there is a chance of 10 times head occurrence and 10 times tail occurrence. The probability for both the head and tail is 50-50 chances. Thus, in flipping the coin 20 times there are 10 chances for head and 10 chances for tail and this is probable. The logical definition of probability as stated by Guilford (1978) is that it is the ratio of the number of ways in which the favoured way can occur to the total number of times the event can occur.

This probability may be also stated in the other way. The probability of occurrence of head in a trial is $\frac{1}{2}$ and the probability of not occurring is $\frac{1}{2}$. The probability of occurrence and not occurrence therefore, constitute 1.00. This may be stated in a formula. The probability of occurrence is stated as p and the non-occurrence is stated as $1-p$ which is represented by the letter q. To state it in another way, $1-p=q$ or $p+q=1.00$.

The probability of obtaining a head in one flipping is $\frac{1}{2}$. What is the probability of obtaining head in two flippings? Can we get two heads in two flippings? Here also the chances are 50-50 in each flipping. If there are two flippings then there may be chances for two heads or no heads or two tails. In this there are four possibilities such as HH, HT, TH and TT. There can be two heads, one head and one tail, one tail and one head and two tails. Therefore, the probability of occurrence of head in two flippings is $\frac{1}{4}$ only. The two flippings are independent and one is not in any way influencing the other. Similarly, if there are three flippings the probability of obtaining 3 heads is $\frac{1}{8}$. This means that there are eight outcomes such as HHH, HHT, HTH, HTT, THH, THT, TTH, TTT or $(½)(½)(½) = \frac{1}{8}$. In the same fashion when the number of trials increase the probability of occurrence of head is getting reduced. Out of possible eight outcomes or events 1 of which is favouring 3 heads, three of the events are favouring 2 heads and 3 of the events are favouring one head and one of the events is favouring no heads.

In flipping or tossing a coin if one event is occurred then the other events do not occur. For example, if there is head in flipping there will not be tail. Therefore, the probability of HT in two flippings is $\frac{1}{4}$. The other two outcomes are HH and TT which have the chances of $\frac{1}{4}$ and $\frac{1}{4}$. This is the probability of occurrence of Head and Tail in four outcomes $\frac{1}{4}+\frac{1}{4}+\frac{1}{4}+\frac{1}{4} = 1.00$.

Suppose, we are tossing a coin four times or toss 4 coins. What are the probabilities of obtaining heads? The chances are (½)) (½) (½) (½) $= \frac{1}{16}$. We can obtain 4 heads only once in 16 times. The following are the possible combinations:

Coin-1	H	H	H	H	H	H	H	H	T	T	T	T	T	T	T	T
Coin-2	H	H	H	T	T	H	T	T	H	H	T	H	H	T	T	T
Coin-3	H	H	T	H	T	T	H	T	H	T	H	H	T	H	T	T
Coin-4	H	T	H	H	H	T	T	T	H	H	H	T	T	T	H	T
	4	3	3	3	2	2	2	1	3	2	2	2	1	1	1	0

The probabilities are 4 heads $\frac{1}{16}$, 3 heads $\frac{4}{16}$, 2 heads $\frac{6}{16}$, 1 head $\frac{4}{16}$ and 0 head $\frac{1}{16}$.

The probability can be illustrated and defined through another example. Let us select 60 red marbles and 40 blue marbles and put them in a small box and mix them well. Let us draw from the box one marble at random. There is a chance of drawing the 60 red marbles out of 100 and 40 blue marbles out of 100 or the proportion of red marbles is 0.60 and the blue marbles is 0.40. Let us assume that we are making 1000 random draws and each drawn marble is replaced in the box again and remixed every time. At the end of 1000 random draws the percentage of red marbles drawn would be 60 and the percentage of blue marbles drawn would be 40. This is how the probability is functioning in the occurrence of events.

What is Chance?

We have been using the concept 'chance' in several places. This needs some explanation. For example, in tossing a coin our assumption is that there is a probability of obtaining the heads half of the times and not heads half of the times. There are two possible alternative events in tossing a coin, head or not head or tail. But what justification of assumption is there that heads are appearing just as likely as the tails are appearing and no more and no less. The answer given to this is that it is 'chance'. According to Hagood (1941) "chance is the result of a multiplicity of simultaneously operating independent factor." We cannot specifically state what factor causes the head and what causes the tail.

Scientists while explaining the probability use this concept invariably and state that is by chance. The meaning of chance in the common sense world, is the way things happen without apparent cause. It is an unpredictable event occurring accidentally. Chance may be defined as an event caused by some unknown unpredictable element resulting in a certain way rather than in another way. Chance also carries the meaning of probability.

BINOMIAL DISTRIBUTION

What is Binomial Distribution?

The binomial distribution is a mathematical model. Blaise Pascar worked out and evolved this model in gambling situation for his friends. This model solves the problem involved in flipping of a coin for a thousand times. The procedures of determining the possible combinations for 3 coins and 4 coins are explained above. There are 8 combinations for 3 coins or 3 flippings of a coin $\left(\frac{1}{2}\times\frac{1}{2}\times\frac{1}{2}\right)$ and there are 16 combinations for 4 coins or 4 flippings of a coin $\left(\frac{1}{2}\times\frac{1}{2}\times\frac{1}{2}\times\frac{1}{2}\right)$. The procedure of determining the possible combination is very laborious if there are 10 coins or 10 flippings of a coin. The easiest thing is to use binomial distribution model. It is a mathematical expression mostly used in statistical applications. In a considerable number of tosses of 2 coins, since there are related results, the probabilities are multiplied. There is a general mathematical expression for this relationship. This can give expression to any number of combinations. This mathematical expression is called the expansion of the binomial and stated in the form of $(p+q)^n$. Therefore, binomial distribution is a statistical distribution giving the probability of obtaining a specified number of successes in a specified number of trials of an experiment, with a constant probability of success in each trial and also the failure. Binomial is a mathematical term, which means two. That which is consisted of two terms and joined by a + is called binomial. The binomial formula is stated in two terms as $(a+b)^n$ or $(p+q)^n$. The name binomial has come because of these two terms.

In the binomial expansion several probabilities would be obtained by the coefficients. The simple equation is $(a+b)^n$. In the case of 2 coins, n refers to the coins. The equation may be stated as: $a^2 + 2ab + b^2$ or 1, 2, 1. In the case of 3 coins independently tossed the equation is: $a^3 + 3a^2b + 3ab^2 + b^2$ or 1, 3, 3, 1. When 4 coins independently tossed the equation extends in the same way: $a^4 + 4a^2b + 6ab + 4ab^2 + b^2$ or 1, 4, 6, 4, 1. The general expansion of the equation $(a + b)^n$ begins with a^n and ends with b^n. In the equation if n is 10 then the multiplied results or the coefficients are 1, 10, 45, 120, 210, 252, 210, 120, 45, 10, 1. The sum of these coefficients is 1024 or 2 to the power of 10 (2^{10}).

This means that the probability of obtaining the heads is $\frac{1}{1024}$ when 10 coins are tossed independently. The probability of obtaining 9 heads is $\frac{10}{1024}$, the probability of 8 heads is $\frac{45}{1024}$, the probability for 7 heads is $\frac{120}{1024}$, for 6 heads $\frac{210}{1024}$, for 5 heads $\frac{252}{1024}$, for 4 heads $\frac{210}{1024}$, for 3 heads $\frac{120}{1024}$, for 2 heads $\frac{45}{1024}$, and for one head $\frac{10}{1024}$ and for no head $\frac{1}{1024}$.

These data can be used to draw a frequency polygon which may have the bell shaped curve or normal probability curve. *The Pascal Triangle* prepared for 10 coins gives the shape of a normal curve (Guilford 1978).

The Use of Subscript Notation in Binomial Expansion

Another simple and more useful way of knowing the binomial expansion is to use p and q instead of a and b. This subscript notation looks more meaningful than a and b. The p represents the probability of obtaining heads or success and q represents the probability of not obtaining the head or failure on a single coin. The equation is: $1 - p = q$ or $q = 1 - p$. Therefore, we have $(p + q)^n$. Let us assume that the n is 2. In tossing a coin if p is $= \frac{1}{2}$, then the equation is $(½)^2 + 2(½)\,(½) + (½)^2$. This may be stated as: $\frac{1}{4}$, $\frac{2}{4}$ and $\frac{1}{4}$. One advantage of p and q notation is that we expect the p to be equal to $\frac{1}{2}$. When the sample size is too big the binomial distribution method can be applied for estimating values and it is more convenient.

The Pascal Triangle is prepared to illustrate the distribution of results and it has a shape of normal curve. This means the results are evenly distributed. But the results or the probabilities are not always normal or symmetrical type. For example, if the p is $= \frac{7}{8}$ and the q is $= \frac{1}{8}$ then the curve is positively skewed. The mean is drawn towards right. If the p is extremely small then the curve is negatively skewed. The mean is pushed towards left end. The binomial expansion offers the probabilities and theoretically expected frequencies for any given n's, p's and q's. The expected frequencies are the number of times the coins tossed. The central value, variation and skewness can be described with such theoretical distributions. The formulae are:

Mean (M) = np

Standard deviation (SD) = $\sqrt{npq}$

Skewness $gl = \frac{q-p}{\sqrt{npq}}$

p = Probability of favoured outcome

$q = 1 - p$

n = number of occurrence of the event

If $p = \frac{1}{2}$ then q is also $\frac{1}{2}$ and, therefore, the mean is not disturbed because the skewness is zero. The level of skewness depends upon the deviation of the p from its probability of $\frac{1}{2}$. The deviation on either side from the probability of $\frac{1}{2}$ may result in skewness. The skewness is positive if the mean is drawn towards right and the skewness is negative if the mean is drawn towards the left end. The skewness and Kurtosis are explained elsewhere in this treatise.

We can describe here the expected frequencies in tossing a coin 10 times or tossing 10 coins independently. Here the $p = \frac{1}{2}$ and $n = 10$. According to the formula the mean = $np = \frac{1}{2}(10) = 5.0$. The standard deviation according to the formula is : $\sqrt{npq}$.

$$SD = \sqrt{10(0.5)\,(0.5)} = \sqrt{2.5} = 1.58$$

The skewness can be worked out by using the formula to a data obtained from tossing 10 coins and if there is any departure from the normal distribution.

Figurative Illustration

A figurative illustration can be made to the binomial distribution. A histogram and a curve can be plotted making use of the expected number of frequencies and of the number of heads by tossing a coin 10 times or by tossing 10 independent coins. The sum of tossing is 1024 times for 10 coins. The number of heads can be put on the *X* axis (ordinate) and the number of expected frequencies on the *Y* axis (abscissa). The number of expected frequencies for 10 coins and the 11 categories of scores are presented above. Here, the score is the number of heads expected in tossing. For 10 coins the histogram and the curve can be drawn. But if the *n* is 20 or 25 or 50 the number of frequencies become larger and larger and the width of distribution also may extend according to the number. In such cases the distribution of width can be retained the same as for the width of 10 coins. If the width is retained the size of the bars of the histogram become thinner and thinner and ultimately they merge with one another and may have the look of a normal curve. This is illustrated in *Fig. 9.1.*

The general principles of binomial distribution $(p+q)^n$ can be expanded. Each term in binomial expansion indicates the number of correct items of successes (p) and the number of incorrect items of failures (q) and the coefficients. The coefficients are obtained in a number of ways (n), the combinations of successes (p) and failures (q) have occurred. Therefore, each term or cell of the binomial expansion is the product of the coefficient, the power of p and the power of $q(p+q)^n$. How the coefficients are estimated? It is explained here. The coefficients are frequencies in a sense. They are the expected number of times the heads may fall in a total of 1024 tosses.

In the binomial expansion the first term or cell is 1. This is a coefficient in numerical value. The power of p in the first cell is always n and the power of q is always zero. Since it is zero it is not mentioned and, therefore, the first term or cell is *pn*. The coefficient value of the first cell is, therefore, obtained by using the formula $p^n = (0.5)^{10} =$ 0.0009765 or $\frac{1}{1024}$ the formula $(p+q)^n$ is stated as p^n or $(0.5)^{10}$ as the q is zero and the n is 10. The result is 0.0009765 or $\frac{1}{1024}$. That is, in 1024 tosses of the 10 coins there is a chance of obtaining all 10 coins as the heads only once.

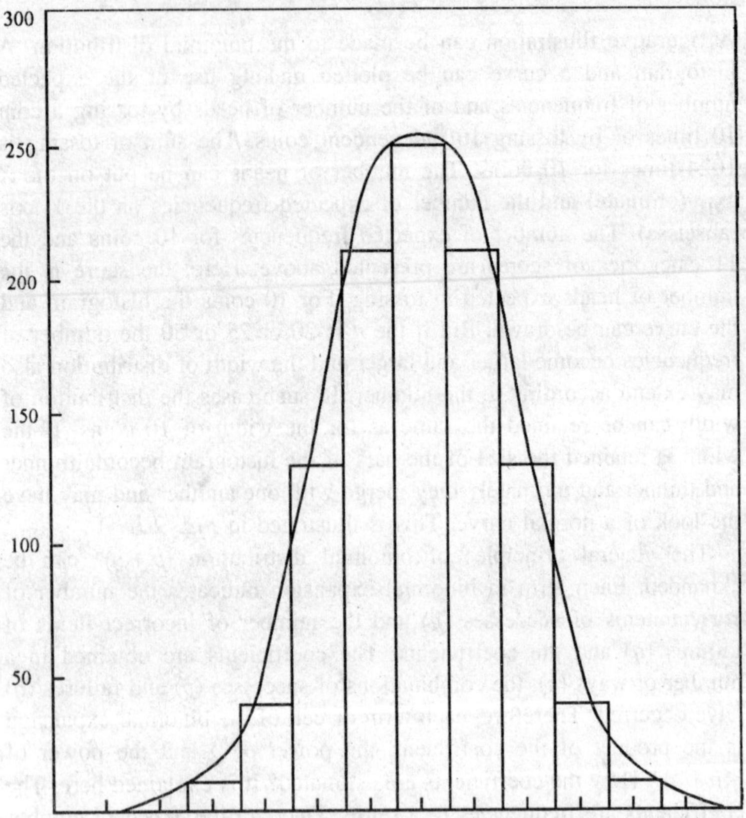

Fig. 9.1 : Histogram and Normal curve fitted to Binomial distribution

In the ensuing every term or cells the power of p is decreasing by one in a systematic way and the power of q is increasing by one in a systematic way. This process goes on till the q gets the power of n. The value of the coefficient in any cell and the power of p are multiplied and divided by the power of q plus one $(q^n + 1)$ in the particular cell. This will give the numerical value of the coefficient. This can be worked out in the same manner as shown above. For example, the numerical coefficient 210 of the 5th term or cell is obtained by multiplying the coefficient of the 4th term or cell by its corresponding power of p (7) and then dividing it by the power of q (3) plus 1. Thus we get

$$\frac{p^7}{q^{3+1}} = \frac{(120)(7)}{3+1} = \frac{840}{4} = 210$$

Though the above example seems to be simple there are certain intricacies in evolving certain coefficients. Therefore, it is better to show how all the coefficients or frequencies of falling heads are worked out for all 10 coins tossed 1024 times.

Table 9.1 : Showing the p value, coefficients and q value in tossing 10 coins 1024 times

Coins : p	10	9	8	7	6	5	4	3	2	1	0
Coefficient:	1	10	45	120	210	252	210	120	45	10	1
q:	0	1	2	3	4	5	6	7	8	9	10

The formula for the first coefficient is : p^{10}, p value is 0.5, n is 10, p to the power of 10 = $(0.5)^{10}$ since q is 0, it is not included in the formula. Therefore, $(0.5)^{10} = 0.0009765$ or $\frac{1}{1024} = 1$. This 1 indicates that when all the 10 coins tossed 1024 times there is a probability of obtaining only once all 10 coins will be falling with heads. The second coefficient is 10. The formula is $\frac{1 \times p}{q+1} : \frac{1 \times 10}{0+1}$. The 1 is the first coefficient, 10 indicates the number of coins, q is 0 since p is 10 and 1 is to be added to q. Therefore, the coefficient is $\frac{1 \times 10}{(0)+1} = \frac{10}{1} = 10$. This means that out of 1024 tosses only 10 times 9 coins will fall with heads. The third coefficient is 45. This is evolved by multiplying the second coefficient by its corresponding $n(9)$ and dividing the sum by q +1, since the p is 9, the q is 1. Therefore, the formula is $\frac{10(9)}{1+1} = \frac{90}{2} = 45$. This shows that 45 times 8 coins will fall with heads out of 1024 tosses.

The same procedure is followed for coefficient 120, using the previous coefficient value and its corresponding $n(8)$. Now the q value is 2, hence 2 + 1 = 3. Therefore, we get $\frac{45(8)}{2+1} = \frac{360}{3} = 120$. This shows that out of 1024 tosses 120 times there is probability of getting 7 coins as heads.

For obtaining the coefficient 210, use the previous coefficient value 120 and its corresponding $n(7)$. The q value is 3. Therefore, the coefficient is $\frac{120(7)}{3+1} = \frac{840}{4} = 210$. In this case 6 coins out of 1024 tosses will fall with heads 210 times.

The coefficient 252 is obtained by using the coefficient 210 and its n value 6 and dividing the sum by the q value 4 and plus one $\frac{210(6)}{4+1} = \frac{1260}{5} = 252$. Out of 1024 tosses there is a probability of getting heads from 5 coins 252 times.

In the same way we estimate the next coefficient value 210. The n value is 5 and the q value is 5 and the plus one is constant. Thus, we get: $\frac{252(5)}{5+1} = \frac{1260}{6} = 210$. Out of 1024 tosses 4 coins will fall with heads 210 times.

The coefficient of 120 is obtained by making use of 210 and multiplying it by its corresponding n value 4 and dividing it by the q value 6 and the constant 1.

Thus, we have $\frac{210(4)}{6+1} = \frac{840}{7} = 120$. There is a probability of getting heads 120 times from 3 coins out of 1024 tosses.

The next coefficient value is 45. This is obtained by multiplying the previous coefficient value 120 by its corresponding n value 3 and dividing the sum by 8 (q value 7 and the constant 1).

Thus, $\frac{120(3)}{7+1} = \frac{360}{8} = 45$. The probability of getting all heads from 2 coins is now 45 times out of 1024 tosses as the p value is decreased and the q value is increased.

The coefficient 10 is obtained by multiplying the previous coefficient value 45 by its n value 2 and dividing the sum by the q value 8 and plus one $\frac{45(2)}{8+1} = \frac{90}{9} = 10$. Out of 1024 tosses there is a chance of getting all heads from ten coins only 10 times.

The last coefficient value is 1. The p value or n is 1 and the q value or n is 9. Applying the same procedure we obtain the coefficient value : $\frac{10(p)}{(q)+1} = \frac{10(1)}{9+1} = \frac{10}{10} = 1$. This means that out of 1024 tosses there is a probability of getting no head from all 10 coins only once. This is indicated by 0 in the Table 9.1.

In this way, tossing 10 coins 1024 times we get all heads only once from 10 coins, for 9 coins 10 heads in 1024 times, for 8 coins 45 heads in 1024 times, 7 coins have 120 heads in 1024 times, 6 coins have 210 heads in 1024 times, 5 coins have 252 heads in 1024 times, 2 coins have 45 heads in 1024 times, one coin have 10 heads in 1024 times and out of 1024 tosses only once there will be no head from all the coins. This may be concluded like this. All the 10 coins will fall with heads only once in 1024 tosses and all the 10 coins will fall with no heads only once in 1024 tosses. This is the crux of the binomial distribution.

10

TESTING HYPOTHESIS

One Tailed and Two Tailed and Two Types of Errors

Introduction

In social science researches, the researchers show more interest in two important aspects. One is the formulation of hypotheses and the other is the testing of the hypotheses. A problem exists and the researcher is interested in getting it tested. Hence, he is devising practical solutions for the problems. But in the fields of science facts and events are justified and proved by evolving logical conclusions. Hypotheses were evolved in the context of discoveries. Several scientific inventions were accidental and were not based upon any specific hypotheses. For example the famous Archimede's principles. Archimedes, the Greek mathematician and physicist who lived in 3rd century BC was able to discover that the apparent loss in weight of a body immersed in a fluid is equal to the weight of the displaced fluid accidentally when he jumped into the bath tub. Several times he might have took bath in a tub. But only that particular incident made him to discover this principle. The British Physicist Isaac Newton's Law of Gravitation resulted from the incident of falling an apple on his head. The invention of steam engine by James Watt happened accidentally while he was observing the boiling water throwing away the lid of the vessel. Similarly, the invention of magnetic field by Ampere, the French Physicist. These are some of the accidental events that resulted in scientific inventions and discoveries. These were not based upon any hypotheses and testing. The scientists later on reconstruct those thought processes in a logical manner and systematically present them in a scientific report. This is what McGuigan (1978) observes. According to him when a scientist states a hypothesis and the material related to it, he does not relate how the hypothesis is actually arrived at. In the above quoted inventions and discoveries this can be observed. But we cannot say that all scientific inventions and discoveries were not based upon hypotheses.

In social science, particularly in psychological research, the psychologist observes the stimulus response events and formulates his hypothesis. Or it may be for verification of some existing theory or laws or he may want to repeat the same experiment. There are several definitions for a hypothesis. A hypothesis is a tentative solution to a problem. Black and Champion (1976) state that a hypothesis is a tentative statement about something, the validity of it is not known. According to McGuigan (1969) a hypothesis is a meaningful statement of the potential relationship between two or more variables. In other words a hypothesis is a statement of tentative solution for the problem in study involving the relationship between the independent variable and dependent variable. In a testing which is designed to find out the solution to the problem there can be any number of independent and dependent variables and also the results.

Testing Hypothesis

In testing a hypothesis, formulated by the researcher, there are certain prerequisites the researcher should adapt. The hypotheses should be subjected to some sort of empirical examination and decide whether they would be accepted or rejected by the observations or the results of the study. Therefore, the primary requisite is a real and suitable social situation in which the hypothesis is tested without encountering much difficulty in collecting the samples. The other prerequisite is that the researcher should see that his hypotheses are testable. The variables used in the hypotheses should be suitable for measurement of some kind.

After having collected the data towards testing the hypotheses, the researcher treats his data statistically and presents them in tabular columns. How, what are the bases upon which he can draw the conclusions and state that his hypotheses are accepted or rejected. There is every possibility for the researcher to be subjective in his interpretation of the results. In other words, he should have taken some guidance on this subject since he is setting up boundary lines between the critical and non-critical regions. There are certain decisions which should be taken in advance in testing the hypotheses. Statisticians often insist upon this, which is called *rules for statistical decisions* or *decision rules*. This is one way of specifying the conditions. The researcher should decide whether he is accepting or rejecting the hypotheses, if so what are the significant levels to adopt or to reject them. Hypotheses, whether statistical or null or research are generally concerned with differences and deviations or errors.

Therefore, it is better to delineate the general problem of hypothesis testing and discuss the problem of what level of significance should be adopted following the *decision rules*. In this context the problem of one tailed and two tailed tests are also involved. If several samplings are drawn one by one from a population several statistical values can be derived from them such as frequency distribution, means and other dispersions, the difference between means and critical ratios, etc. Hypotheses are generally concerned with differences. Hypotheses are formulated on the basis of assumptions or prior observations. Every hypothesis set up has a logical background. If hypotheses are set on these basis then they usually include predictions from the experimental results. For evaluating the research hypotheses a null hypotheses may be set up.

While we are estimating the possible differences from the means of the experiment the null hypothesis H_0 (Symbolized as H_0) is set in competition with an alternative hypothesis H_1 (Symbolized as H_1). Suppose, for example, the hypothesis H_0 states that two mean values do not differ ($m_1 = m_2$). The alternative hypothesis H_1 states that the first mean is greater than the second mean ($m_1 > m_2$) or the first mean is smaller than the second mean ($m_1 < m_2$). The researcher is interested in the difference between the two means irrespective of the direction of the difference. Further the hypothesis H_1 also states that the first mean is not equal to the second mean ($m_1 \neq m_2$). Among these which of the alternates is appropriate and to be tested by the researcher.

For example, an experiment was carried out and two samples were collected and the mean values was estimated as m_1 and m_2. The difference between m_1 and m_2 was used to test the hypothesis H_0 in competition with the hypothesis H_1. This means that we are making a decision on the basis of the obtained difference between m_1 and m_2 which hypothesis is true, whether H_0 or H_1. Suppose, for example, if H_0 is true we must state the probability level. Here the appropriate null hypothesis is that $m_1 = m_2$. Let us have any one levels of significance from the levels $p = 0.10$, $p = 0.05$, $p = 0.01$ which are conventionally used. Let us assume that the level of significance chosen in this case is $p = 0.05$.

In this experiment, the obtained result is as small as the chosen level of significance, $p = 0.05$ and therefore, we reject the null hypothesis H_0. This implies the acceptance of the alternate hypothesis H_1. If the result would not have reached the level of significance 0.05 we would have accepted the hypothesis H_0. The acceptance of Hypothesis H_1 means that H_0 is not true and m_1 is either bigger than m_2 or smaller than m_2 ($m_1 > m_2$ or $m_1 < m_2$).

One Tailed and Two Tailed Tests

We have stated above that there could be three possible alternates to do with hypotheses H_0 and H_1. If the hypothesis H_1 states that m_1 is not equal to m_2, a two tailed test takes place. If the hypothesis H_1 states that the sample mean is larger, then one tailed test takes place. Whether we will have to use one tailed test or two tailed test depends upon the practical decision made. It should be observed that whether the hypothesis tested will be having the chance of deviating in one direction or in both directions. For example, if the performance is better than chance then the deviation may be in one direction. In any performance if the participant, is successful in less than (0.5) 50 per cent of the trials the hypothesis may be accepted since the participant cannot perform and show a better discrimination than chance. This means that the participant's performance in discriminating right and wrong or true and false is the same as chance. In such case one tailed test is chosen.

Suppose, for example, in the experimental situation we decide to split the sample 50-50 on some issue then we will have to consider the chance of deviations on both samples or directions. We will have to examine the chance deviations in both the samples. In this case we are using the *two tailed test*.

Another example can be used to explain this phenomenon. There are two groups to be compared on a given problem. The Hindu group consisted of 200 people and Muslim group consisted 200 people in the experiment. The problem is family planning programme. The proportions of acceptance of the programme for both groups are estimated and the difference between the two groups is tested to decide whether the Hindus and the Muslims differ on the problem. In this case we will have to use two tailed test. The hypothesis H_0 that 'there is no difference between the two groups' is rejected as the difference is obtained in this case and also the probability of occurrence is of the chosen level and the hypothesis H_1 is accepted as the difference between the two groups is found to be reliable.

Suppose we are to estimate the significance of an experimental result in which whether the new detergent washing soap is better in washing than the conventionally used soaps. Here also we are comparing two variables. But the decision is to see that the new detergent soap could exhibit a better quality in washing. The result we observe is in one direction only. Here a one-tailed test is more appropriate. But at the same time if we are testing two types of detergent washing soaps with the view of selecting one of them which is more effective, then we are using a two tailed test.

A one-tailed test can be also applied to verify certain theory or to verify certain obtained result of a previous experiment. The advantage of applying one-tailed test here is that a lower level of critical ratio or '*t*' test can be reached. For example, for the level of significance, p = 0.01, the *CR* or *t* test should be 2.58 for two-tailed test, but for the one tailed test the *CR* or *t* test can be 2.33. Similarly, the level of significance p = 0.05, the *CR* or *t* test should be 1.96 for two-tailed test, but for one-tailed test it can be 1.64. This mean that the *CR* or *t* test for one-tailed test need not be as high as for two-tailed test.

Exponents in this field suggest that in testing hypothesis H_1 or in testing two sample means a one-tailed test can be applied if there is reliable difference in direction predicted. But it should not be taken for granted that it can be done in every case unless the investigator has failed to reach a two tailed test.

Two Types of Errors

The test, whether, it is one-tailed test or two-tailed test the researcher is interested only in getting a reliable and significant *CR* or '*t*' test. There is no use of having a *CR* or '*t*' test which is not significant. It should be significant at some level. It is also true that the non-significant *CR* or '*t*' test or a zero mean difference is also a result from an experiment to be discussed statistically. But how big the *CR* or '*t*' test from a measurement should be if it is to be significant. It is the size of the *CR* or '*t*' test that decides the level of significance, whether the *CR* or '*t*' test is significant at 0.05 level or 0.01 level or 0.001 level.

If the observed mean difference is not reaching the level of rejection of null hypothesis then we accept it. But we must make it a point that the rejection of null hypothesis does not mean that there is no difference between the means of two samples. Unless the difference is zero there is some difference.

Psychologists for a long time have been demanding a *CR* of 3.00 to have the difference declared as significant at 0.003 level for two-tailed tests. But if the *CR* did not reach this level (3.00) then we will have to reject the null hypothesis on the ground that the observed mean difference is not having the *CR* 3.00. If this would be the case we are often making errors in our conclusions. For example, if the observed difference between the means of two samples yielded a *CR* or *'t'* test of 0.854 it does not prove that there is no difference. According to Table IV (Appendix B) of Fisher it is significant at 0.4 level. However, smaller the *CR* or *'t'* a significant level for the observed difference has been worked out by Fisher. The observed mean difference yielded a *CR* or *'t'* 0.854 which is significant at 0.4 per cent level. But this difference could be a chance difference in the sample, a deviation from the real difference. This level of difference only indicates that the real difference could be zero. Therefore, we cannot accept that there is real difference in the observed values and state that it is reliable. With this *CR* or *'t'* test we can reject the null hypothesis, but cannot prove that the difference is real. Therefore, in order to reject the null hypothesis we expect a greater *CR* or *'t'* test to have a higher level of significance. And also, the *CR* or *'t'* test need not be ordinarily be as high as 3.00 to yield 0.003 level of significance. Though there are controversies among the psychologists with regard to the level of significance of the observed means there is a general acceptance among them to come down from the rigid level of *CR* 3.00 to a laxed level of 2.58 and 1.96 to have the level of significance at 0.01 and 0.05 levels respectively for the two-tailed tests.

The researcher even after adopting the decision rules is likely to have an opportunity to be bias or quite subjective. But he is bound by certain professional ethics. What does the researcher do in adopting the decision rules? First of all he has to state the level of significance of the result of his proposed study and secondly he has to specify the sampling distribution.

Every hypothesis has two possible conclusions. The hypothesis is either true or false. But there cannot be both true and false at the same time.

This principle is applicable to null hypothesis, statistical hypothesis and to research hypothesis. Already mention has been made with regard to the choice of significant level. This is a risk the experimenter faces in making a statistical decision to accept or reject the hypothesis. At this stage he may be wrong, because he is rejecting the hypothesis on the basis of statistical decision taken earlier. Let us assume that the null H_0 or statistical hypothesis is tested and it is true, but the result yielded an erroneous conclusion and the hypothesis is rejected. Thus, he has committed an error called *type I error.* The type I error is also called Alpha error.

But if the null hypothesis H_0 is false in fact and the test of significance has yielded a result in such a way that the researcher has decided not to reject it and decided to accept on the basis of the level of significance he has committed an error called *type II error*. The type II error is also called *Beta error*.

The researcher generally wanted to eliminate both Type I and Type II errors. But he cannot eliminate them. He can take steps only to reduce these errors. From the foregoing discussions we can understand that the type I error is under the direct control of the researcher and therefore, there is a possibility of reducing it. But he can control the type II error indirectly. Any attempt to control one type of error also causes change in another type of error. The relationship between these two types of errors is on the inverse direction. If one is reduced, the other is increased and *vice versa*.

It becomes necessary to clarify the process of choosing a level of significance according to the decision rules. The researcher can set the probability level for rejecting the true hypothesis. It is indicated by (a), alpha or class of result for rejecting the null hypothesis H_0. An alpha can be 0.05, or 0.01 or 0.001. The researcher makes the decision to reject the hypothesis if the result occurs within this class or alpha which he has already fixed. Therefore, an alpha of 0.05 means that when the hypothesis is true there are 5 chances in a 100 for rejecting the null hypothesis. He is wrong 5 times in a 100 and the frequency of type I error is 5 in 100. Hence, the probability of type I error is 0.05. The researcher by choosing an alpha smaller than 0.05, to a class of results that would occur less than 5 per cent of times say, only one chance in a 100 when the null hypothesis is true, the frequency of error can be reduced in type I errors. In this case there is only one chance of being wrong. By choosing an alpha or a class of results 0.001 the probability of type I error is 0.001. But one unpleasant occurrence is that while reducing the chances of type I errors by choosing a smaller alpha the frequency of type II errors is increased. This is in accordance with the statement we made already that the type I errors is having inverse relationship with type II errors. Hoel (1947), Johnson (1949) and Mood (1950) have discussed Type I and Type II errors in greater detail and state that under certain circumstances type I error is more serious than type II error and under some other circumstances, type II error is having more serious effect than type I error. According to them if we expect with a smaller probability for rejecting the null hypothesis there is likelihood of increasing type II errors and we may fail to reject the null hypothesis which is not true.

11

CHI-SQUARE

Introduction

The various statistical treatments to the experimental data discussed in the preceding sections are intended to present the results precisely and accurately so as to make the study more scientific. But each statistic can provide only a limited information which is not sufficient for our purpose. We need a statistic which can deal with wider area involving more than one set of data to be tested and the level of significance can be estimated. The one statistic that can serve several purposes unlike other statistics is the chi-square. The chi-square can be used to simple data.

The data used in chi-square treatment are in the form of frequencies or that can be transformed into frequencies. The nature of chi-square can be completely explained on the basis of the standard score z. The chi-square is equated with χ^2 in the case of one degree of freedom. In this case the formula is:

$$\chi^2 = Z^2 = \frac{(X - \bar{X}\ )^2}{SD^2 \text{ or } \sigma^2}$$

Where X = score obtained in a population

$\bar{X}$ = the mean score

SD or σ = Standard deviation

The '*t*' Test, *z* and Chi-Square

While discussing '*t*' test we have stated that in a frequency or sampling distribution the *t* ratio is the same as of *z*. The *z* is equated with critical ratio. Since the difference between critical ratio and *t* ratio is only with regard to the size of the sample, the z is identified with 't'. Here again the Chi-square is identified with Z^2. It is because the chi-square is also used as a sampling statistic like z. Therefore, both the chi-square and z has sampling distribution. But in this case of z it is normal whereas in the case of chi-square it is not normal as the chi-square is more directly related to z^2. The relation the chi-square has with z is for only one degrcc of freedom. But with chi-square several hypotheses need to be tested and hence, there are more than one degree of freedom. The one degree of freedom can be illustrated with an example.

Experimental Illustration

An investigation, for example, is conducted among 50 postgraduate students in a university, whether university students can participate in political activities. Among the 50 who were questioned, 35 have given affirmative answers and 15 have given negative answers. If the answers are equally divided in a population then the possible frequencies could be 25-25. Then the null hypothesis in this case should be that there is no difference between the mean of those who agree the idea and those who disagree the idea. Then the expected frequencies in a sample of 50 is = Np. $50 \times 0.5 = 25$. Therefore, it is assumed that the mean of the population distribution of frequencies is 25.

According to the chi-square formula the mean is called the expected frequency *fe*. The obtained frequency can be compared with *fo*. Is the obtained frequency of 35 differs significantly from the expected frequency of 25, which is expected in the null hypothesis.

The formula to test the hypothesis is: $Z = \frac{fo - fe}{\sqrt{Npq}}$

Where fo = obtained frequency

fe = expected frequency

N = Number of cases

p = 0.5 proportion

$q = 1 - 0.5$

$$z = \frac{35 - 25}{\sqrt{50 \times 0.5 \times 0.5}} = \frac{10}{\sqrt{12.5}} = \frac{10}{3.54} = 2.82$$

From the 't' ratio table we can find out that the z value 2.82 is higher than the table value 2.58 for 0.01 level significance. Therefore, the difference is significant at one per cent level. In this equation the mean is 25 and the variance is $50 \times 0.5 \times 0.5 = 3.54$.

Chi-Square and Z

Now let us test the chi-square hypothesis with the same data.

The formula for two frequencies = $X^2 = Z^2 = \frac{(fo - fe)^2}{Npq}$

$$\text{or} \qquad \chi^2 = \frac{2(fo - fe)^2}{fe} \qquad \text{......(Eqn. 2)}$$

In this equation where $p = q = 0.5$

The Npq can be stated as $\frac{fe}{2}$

$$fe = NP, Q = 0.5 \text{ or } 1/2$$

$$\chi^2 = \frac{2(35 - 25)^2}{25} = \frac{2(10)^2}{25} = \frac{200}{25} = 8.0$$

From the chi-square table it is found that the χ^2, 8.0 is little higher than the table value 6.635 and hence is significant at 0.01 level. This means that if 100 investigations are conducted 99 times we have the chance of getting the same result. Only once the result may be different. Since we equated the X^2 with z^2, the root of X^2 will give us the z value. Thus, the root of X^2 = 8 = 2.82 = z. This is the value we have already obtained in equation 2. If the degree of freedom is 1 then the Z and X^2 are the same. According to equation 3, $X^2 = Z^2$. If we want to determine the Z value then we will have to find out the root of chi-square Z^2. But if the degrees of freedom are more than one we cannot equate chi-square with Z value.

Chi-Square test can be used to test the hypothesis by combining more than one test or test scores with more than two responses.

Experimental Illustration

In another experiment on the same problem there were 48 students of both men and women students of equal number and the responses were 'Yes', 'No' and 'Undecided'. The problem is whether there is any genuine sex difference among the students in the view of participating in political activities. Therefore, the null hypothesis is that there is no sex difference in all the three frequencies taken together. The formula applied to estimate the chi-square is:

$$\chi^2 = \Sigma \frac{(fo - fe)^2}{fe}$$

Table 11.1 : Showing the expected frequencies and obtained frequencies and the chi-square (*N*-48)

Response	*fe*			*fo*		*fo − fe*		$(fo-fe)^2$		$\frac{(fo-fe)^2}{fe}$		
	M	*W*	*Both*	*M*	*W*	*M*	*W*	*M*	*W*	*M*	*W*	*Both*
Yes	8	8	16	10	6	+2	−2	4	4	0.5	0.5	1.0
No	8	8	16	4	12	−4	+4	16	16	2.0	2.0	4.0
Undecided	8	8	16	10	6	+2	−2	4	4	0.5	0.5	1.0
Sum	24	24	48	24	24	0	0			3.00	3.00	$6.00 = X^2$

M = Men; *W* = Women

The chi-square table figure 5.991 is the nearest to the obtained figure 6.0

The sample consisted of 48 university students. There were 24 men students and 24 women students. The responses were three. Therefore, the expected frequencies in two groups and for 3 responses should be 24 Men and 24 Women and 8 Yes responses for men, 8 No responses and 8 Undecided responses. Similarly, for 24 women students the same 8+8+8 responses for 'Yes', 'No' and 'Undecided'. But the obtained frequencies are different.

The frequencies are given in Table 11.1. According to the formula the expected frequency (*fe*) should be deducted from the obtained frequency (*fo*). The obtained frequency is the response from both men and women for 3 categories of responses. Therefore, in Table 11.1 the column *fo − fe* should be completed for all men and women and for the 3 responses. In this case men had the frequencies for 3 responses (*fo*) as 10, 4 and 10. The women had the obtained frequency for 3 responses (*fo*) as 6, 12 and 6. For both groups the *fo − fe* are +2, −4, +2 for men, −2, +4, −2 for women. When these values are squared we have 4, 16, 4 for men and 4, 16 and 4 for women. The minus and plus signs were eliminated. These values in columns $(fo-fe)^2$ are divided by the respective *fe*. The values used for dividing $(fo-fe)^2$ are the same, i.e., 8, 8 and 8 for both men and women and also for the 3 responses. Thus, we get under the column $\frac{(fo-fe)^2}{fe}$ the values for each group and for each response category. These values are to be summed up for men and women and for both. The sum of the sums is the chi-square value 6.0. The sum for each response category also is to be made for specific interpretation of the chi-square, i.e. for 'Yes', or 'No' or Undecided. This is given under the column 'both'.

Degrees of Freedom for Chi-Square

Before going to the table to find out the level of significance we must find out the degrees of freedom. The easiest way of determining the degrees of freedom is to count the number of rows and columns and then use the formula $df = (r = 1)\ (c - 1) = (3 - 1)\ (2 - 1) = 2$. Therefore, the *df* is only 2. For the 2 *df* the table figure is 5.991 in the second row which is little smaller than the obtained figure 6.0. This is significant at 0.05 level. The chi-square value 6.0 may occur by random sampling a little less than five times in a hundred times ($P < 0.05$). Therefore, we can reject the null hypothesis that there is no sex difference with significance difference. This means that both the men students and women students have different views regarding the problem and both sexes differ significantly. Looking at the three response frequencies it is observed that the 'No' response also is having the chi-square 4.0 for one *df* which is significant at 0.05 level. From the 'No' answer it appears that women students were decidedly more in favour of not participating in politics. But with regard to undecided response frequencies are concerned more men students were not able to decide the matter.

12

DUNCAN'S RANGE TEST

A Multi-group Design

Introduction

Psychologists are very much interested in using more number of groups, particularly more than two groups in their experiments and wanted to assess whether these groups really differ from each other. They need several groups to compare, for which they wanted to have the statistics most suitable for this type of study. Psychologists and statisticians provide different statistical procedures. But there is no unified opinion on this matter. Different hypotheses are tested with different methods and some procedures are more emphasised.

Researchers, who are interested in the study of more than two groups meant, the comparison of mean values of the groups. It is not combining two groups and comparing with another two groups. It is only comparing between pairs of two individual groups. It is comparing between any two of several groups at a time in all possible combinations. This can be explained through an example. A researcher has conducted an experiment and collected data from four groups and computed the means for the four groups. The researcher's primary interest is whether group 1 really differs from group 2 since the means vary from group to group. Also wanted to see whether group 1 differs from group 3 and group 4 and whether group 2 differs from group 3 and from group 4. Similarly, whether group 3 differs from group 4. Here the researcher's interest is not to see whether groups 1 and 2 together differ from groups 3 and 4 put together or to see whether groups 1 and 3 together differ from groups 2 and 4 put together.

Statistical Procedures for Multi-group Design

There are several statistical procedures for testing the mean differences between the groups or among the groups:

(1) Among several types of statistical procedures that can be used for more than two randomised group design experiments, psychologists more widely use the procedure of analysis of variance. This is supposed to be called classical statistical method applied to multi-group design.
(2) Another procedure considered to be the most popular test for multi-group comparisons is Scheffe's (1953) method.
(3) A third procedure that is also popular in psychological researches is Ryan's (1959) multiple comparison method.
(4) The fourth is Tukey's (1949) procedure for comparing individual means.
(5) After making use of several available statistical methods, psychologists find that a new multiple range test called Duncan's Range Test is the most suitable for multi-group design experiments.

Duncan's Range Test

In a multi-group experimental design the researcher need not go on testing all the procedures one by one. It is better if he tries the most popular two tests, the Analysis Variance and the Duncan's Range Test (1955). We have described analysis of variance and the factorial design elsewhere. Both are quite suitable for multi-group tests. The recent developments in the application of statistics to psychological research yielded a most convenient method for multi-group testing is the Duncan's Range test. McGuigan (1978) suggests that Duncan's Range Test is most suited for multi-group tests. But it has to be used very carefully as it involves several measures to reach the final stage. There are five steps in computing the 'least significant ranges' and comparing the mean difference. But all are simple procedures.

The important components in the equation of Duncan's Range Test are:

(1) The mean scores of the groups.
(2) The sum of squares of every group. $\left[SS = \Sigma X^2 - \frac{(\Sigma X^2)}{n}\right]$

(3) The square root of the error variance (S_e), $S_e = \sqrt{\dfrac{SS_1 + SS_2 + SS_3}{(3(n-1)}}$, (assuming that there are only 3 groups). The S_e is constant for all the groups.

(4) The degrees of freedom. ($df = N - r$), N = number of participants in the experiment and r = number of groups.

(5) The r_p value. This is the 'least significant standardized range'. This is obtained from the Table of Duncan's Range Test r_p value 5 per cent significant (Table 12.2) for the appropriate degrees of freedom. This is used for working out the important R_p values.

(6) The R_p values. This is the 'least significant range'. This is obtained by the formula: $R_p = S_e\, r_p \sqrt{\dfrac{1}{n}}$. The error variance ($S_e$) and the 'least significant standardized range' (r_p) value and the 'n' of the respective group are involved in estimating the R_p values.

I. The important characteristic of Duncan's Range Test is that the mean difference between groups is compared with the R_p value and the *mean difference should be higher than the respective R_p value, the 'least significant range' for the result to be significant.* Another important characteristic of it is that while comparing the mean differences and the R_p values, the R_p value to be compared should go from right to left and the mean differences should be estimated always between two extreme groups and compared with the R_p values.

The Duncan's Range Test needs to be illustrated with at least 2 or 3 examples as the formula are different for unequal n's.

Duncan's Four Group Design : An Illustration

Suppose, for example a small achievement test was conducted among four groups of 40 students drawn equally and randomly from four schools. In each group there were 10 students. The score of each student was obtained from all four groups and given in Table 12.1.

The first stage of the computation is over, and we got the values of ΣX, ΣX^2, n and $\overline{X}$ for each group. They are given in Table 12.1.

Table 12.1 : Scores of four groups for estimating the levels of mean differences in Duncan's Range Test

S.No.	Group 1		Group 2		Group 3		Group 4	
	X	X^2	X	X^2	X	X^2	X	X^2
1	22	484	20	400	23	529	24	576
2	24	576	23	529	26	676	24	576
3	23	529	24	576	28	784	36	1296
4	20	400	22	484	24	576	34	1156
5	20	400	22	484	25	625	34	1156
6	18	324	20	400	24	576	35	1225
7	19	361	20	400	22	484	28	784
8	20	400	22	484	28	784	37	1369
9	24	576	23	529	29	841	28	784
10	26	676	25	625	30	900	30	900
Σ	216 ΣX	4726 ΣX^2	221 ΣX	4911 ΣX^2	259 ΣX	6775 ΣX^2	310 ΣX	9822 ΣX^2
N	10		10		10		10	
$\bar{X}$	21.6		22.1		25.9		31.0	

II. Then the next stage is to compute the sums of squares for each group. The formula to compute the SS is:

$$SS = \Sigma X^2 - \frac{(\Sigma x)^2}{n} \qquad \ldots (1)$$

Applying this formula to the data obtained in Table 12.1, we have:

$$SS_1 = 4726 - \frac{(216)^2}{10} = 4726 - 4665.6 = 60.4$$

$$SS_2 = 4911 - \frac{(221)^2}{10} = 4911 - 4884.1 = 26.9$$

$$SS_3 = 6775 - \frac{(259)^2}{10} = 6775 - 6708.1 = 66.9$$

$$SS_4 = 9822 - \frac{(310)^2}{10} = 9822 - 9610 = 212.0$$

The next step is to compute the square root of the *error variance* (S_e) for the four groups. The equation is:

$$S_e = \sqrt{\frac{SS_1 + SS_2 + SS_3 + SS_4}{4\,(n-1)}} \qquad \ldots (2)$$

For the computation of the error variance (S_e) for all four groups the procedure is simply to add the squares for all of the groups together and this becomes the numerator of the equation. This equation can be applied to any number of groups. The denominator of this equation is the number of groups multiplied by the number of participants in all the groups put together minus one. We are having four groups in our example and therefore, we have four sums of squares to be added. Since the participants (n) in each group are in the same number, $n_1 = n_2 = n_3 = n_4$ and we can simply multiply n–1 by 4. Applying equation 2 above to the obtained sums of squares given above, we have :

$$S_e = \sqrt{\frac{60.4 + 26.9 + 66.9 + 212.0}{4(10-1)}} = \sqrt{\frac{366.2}{36}} = \sqrt{10.17} = 3.19$$

The error variance for all the four groups is 3.19 (S_e).

III. The next step is to determine the degree of freedom in this test. The formula is: $N - r$.

This capital N refers to all the participants put together in this test and r refers to the number of groups in this test. Therefore, the degrees of freedom here is 40 – 4 = 36.

IV. Now, the next step is to find out the values of r_p and R_p for four groups with 36 degrees of freedom. We do not know whether our results are going to be significant at 5 per cent level or one per cent level. Therefore, let us make use of *Duncan's Range Test Table for 5 per cent* (Table 12.2). Now we can go into the Ratio table for obtaining the values for r_p.

There are four groups here and enter the values in Table 13.3 below. Going down from 1 to 30 and 40 degrees of freedom we get the value of r_4 for 36 *df* 3.11 by interpolation method (Table 12.2). The value for r_3 in the same manner we get 3.03 and for r_2 the value is 2.88. Let us present all these values in a Table and compute the R_p values.

Table 12.3 : The values of r_p and R_p for four groups with 36 *df.*

Values	*Group 2*	*Group 3*	*Group 4*
r_p	2.88	3.03	3.11
R_p	2.91	3.05	3.14

Table 12.2 : Values of r_p for Duncan's Range Test (significance level = 5 per cent)

df	*Number of Groups* 2	3	4	5	6	7	8	9	10	12	14	16	18	20	50	100
1	18.0	18.0	18.0	18.0	18.0	18.0	18.0	18.0	18.0	18.0	18.0	18.0	18.0	18.0	18.0	18.0
2	6.09	6.09	6.09	6.09	6.09	6.09	6.09	6.09	6.09	6.09	6.09	6.09	6.09	6.09	6.09	6.09
3	4.50	4.50	4.50	4.50	4.50	4.50	4.50	4.50	4.50	4.50	4.50	4.50	4.50	4.50	4.50	4.50
4	3.93	4.01	4.02	4.02	4.02	4.02	4.02	4.02	4.02	4.02	4.02	4.02	4.02	4.02	4.02	4.02
5	3.64	3.74	3.79	3.83	3.83	3.83	3.83	3.83	3.83	3.83	3.83	3.83	3.83	3.83	3.83	3.83
6	3.46	3.58	3.64	3.68	3.68	3.68	3.68	3.68	3.68	3.68	3.68	3.68	3.68	3.68	3.68	3.68
7	3.35	3.47	3.54	3.58	3.60	3.61	3.61	3.61	3.61	3.61	3.61	3.61	3.61	3.61	3.61	3.61
8	3.26	3.39	3.47	3.52	3.55	3.56	3.56	3.56	3.56	3.56	3.56	3.56	3.56	3.56	3.56	3.56
9	3.20	3.34	3.41	3.47	3.50	3.52	3.52	3.52	3.52	3.52	3.52	3.52	3.52	3.52	3.52	3.52
10	3.15	3.30	3.37	3.43	3.46	3.47	3.47	3.47	3.47	3.47	3.47	3.47	3.47	3.48	3.48	3.48
11	3.11	3.27	3.35	3.39	3.43	3.44	3.45	3.46	3.46	3.46	3.46	3.46	3.47	3.48	3.48	3.48
12	3.08	3.23	3.33	3.36	3.40	3.42	3.44	3.44	3.46	3.46	3.46	3.46	3.47	3.48	3.48	3.48
13	3.06	3.21	3.30	3.35	3.38	3.41	3.42	3.44	3.45	3.45	3.46	3.46	3.47	3.47	3.47	3,47
14	3.03	3.18	3.27	3.33	3.37	3.39	3.41	3.42	3.44	3.45	3.46	3.46	3.47	3.47	3.47	3.47
15	3.01	3.16	3.25	3.31	3.36	3.38	3.40	3.42	3.43	3.44	3.45	3.46	3.47	3.47	3.47	3.47
16	3.00	3.15	3.23	3.30	3.34	3.37	3.39	3.41	3.43	3.44	3.45	3.46	3.47	3.47	3.47	3.47
17	2.98	3.13	3.22	3.28	3.33	3.36	3.38	3.40	3.42	3.44	3.45	3.46	3.47	3.47	3.47	3.47
18	2.97	3.12	3.21	3,27	3.32	3.35	3.37	3.39	3.41	3.43	3.45	3.46	3.47	3.47	3.47	3.47
19	2.96	3.11	3.19	3.26	3.31	3.35	3.37	3.39	3.41	3.43	3.44	3.46	3.47	3.47	3.47	3.47

(Contd.)

Table 12.2 *(Contd.)*

df	*Number of Groups* 2	3	4	5	6	7	8	9	10	12	14	16	18	20	50	100
20	2.95	3.10	3.18	3.25	3.30	3.34	3.36	3.38	3.40	3.43	3.44	3.46	3.46	3.47	3.47	3.47
22	2.93	3.08	3.17	3.24	3.29	3.32	3.35	3.37	3.39	3.42	3.44	3.45	3.46	3.47	3.47	3.47
24	2.92	3.07	3.15	3.22	3.28	3.31	3.34	3.37	3.38	3.41	3.44	3.45	3.46	3.47	3.47	3.47
26	2.91	3.06	3.14	3.21	3.27	3.30	3.34	3.36	3.38	3.41	3.43	3.45	3.46	3.47	3.47	3.47
28	2.90	3.04	3.13	3.20	3.26	3.30	3.33	3.35	3.37	3.40	3.43	3.45	3.46	3.47	3.47	3.47
30	2.89	3.04	3.12	3.20	3.25	3.29	3.32	3.35	3.37	3.40	3.43	3.44	3.46	3.47	3.47	3.47
40	2.86	3.01	3.10	3.17	3.22	3.27	3.30	3.33	3.35	3.39	3.42	3.44	3.46	3.47	3.47	3.47
60	2.83	2.98	3.08	3.14	3.20	3.24	3.28	3.31	3.33	3.37	3.40	3.43	3.45	3.47	3.48	3.48
100	2.80	2.95	3.05	3.12	3.18	3.22	3.26	3.29	3.32	3.36	3.40	3.42	3.45	3.47	3.53	3.53
∞	2.77	2.92	3.02	3.09	3.15	3.19	3.23	3.26	3.29	3.34	3.38	3.41	3.44	3.47	3.61	3.67

Experimental Psychology, 3/E by McGuigan, ©1978, Adapted by permission of Prentice-Hall, Inc., Upper Saddle River, NJ. p. 245.

The R_p values are obtained by multiplying the r_p values by the error variance S_e and by multiplying the product by $\sqrt{\frac{1}{n}}$. The full equation for obtaining the R_p is: $R_p = Se\ r_p \sqrt{\frac{1}{n}}$.

Where Se = the error variance for all four groups is constant (3.19)

r_p = the least significant standardised ranges obtained from Ratio table for appropriate *dfs* (Table 12.2).

By applying the above equation to each r_p value we get the value

for 2 groups the R_p is $2.88 \times 3.19 \sqrt{\frac{1}{10}} = 2.91$

The value for 3 groups : $R_p = 3.03 \times 3.19 \sqrt{\frac{1}{10}} = 3.05$

The value for 4 groups : $R_p = 3.11 \times 3.10 \sqrt{\frac{1}{10}} = 3.14$

These values are entered in Table 12.3 against R_p under r_p values.

V. The next step is to arrange the means from lowest to highest. Thus we have them from Table 12.1.

Table 12.4 : Means from lowest to highest

Measure	Gr. 1	Gr. 2	Gr. 3	Gr. 4
Mean	21.6	22.1	25.9	31.0

Having ordered the means from the lowest to the highest we must first test the difference between the two extreme means of the four groups. *The extreme means are for groups 4 and 1.* The difference between them is $31.0 - 21.6 = 9.4$. We are considering the extreme values of four groups. The R_p value for group 4 is $R_4 = 3.14$ (Table 12.3). The difference between group 4 and group 1 is 9.4 which is larger than the R_p value 3.14. Hence, we can conclude that *group 4 is significantly different from group 1* at the 5 per cent level.

Next, we can compare the means between *groups 4 and 2.* The difference between the means is $31.0 - 22.1 = 8.9$. The comparison is between two extreme values of three groups. The R_p value is 3.05. The mean difference 8.9 is higher than 3.05. Therefore, *group 4 is reliably different from group 2 at 5 per cent level.*

Lastly groups 4 and 3 can be compared between the means. The difference between the two means is 31.0 – 25.9 = 5.1. This is the comparison between two extreme values of two groups. The R_p value is 2.91. The mean difference 5.1 is higher than R_p value 2.91. Therefore, we can conclude that the difference between groups 4 and 3 is significant at 5 per cent level.

We shall next test the mean difference between groups 3, group 2 and 1. The mean difference between the extreme groups 3 and 1 is 25.9 – 21.6 = 4.3. The R_p value is 3.05 (Table 12.3). The difference between the means of the groups 3 and 1 is higher than the R_p value. Therefore, we can conclude that group 3 differs from group 1 significantly at 5 per cent level.

Similarly, we must test the mean difference between the extreme groups 3 and 2. The mean difference between these two groups is 25.9 – 22.1 = 3.8. The R_p value is 2.91. The mean difference is higher than the R_p value. Therefore, there is difference between groups 3 and 2 and the difference is significant at 5 per cent level.

The next extreme groups to be compared are groups 2 and 1. We are dealing with 2 groups, 2 and 1. The mean difference between extreme groups 2 and 1 is 22.1 – 21.6 = 0.5. The mean difference is very low. The R_p value for the two groups is 2.91. The mean difference is lower than the R_p value. Therefore, we can conclude that the mean difference between groups 2 and 1 is not significant.

The mean difference between group 4 and other groups were tested and group 4 is found to be reliably significant from groups 3, 2 and 1 at 5 per cent level. The mean for group 4 is higher than groups 3, 2 and 1. Therefore, we can conclude that group 4 is significantly superior to other groups. A scheme is evolved to indicate the fact. *Any two means that are underscored by the same line are not significantly different. Any two means that are not underscored by the same line are significantly different.* Hence for group 4 the significance is indicated by not drawing a common line.

	Gr. 1	*Gr. 2*	*Gr. 3*	*Gr. 4*	
Mean	21.6	22.1	25.9	31.0	(Group 4 compared)

The mean differences between group 3 and groups 2 and 1 are tested and group 3 is found to be significantly different from groups 2 and 1. We can conclude that group 3 is reliably superior to groups 2 and 1. Therefore a common line is not drawn to indicate the result.

	Gr. 1	*Gr. 2*	*Gr. 3*	*Gr. 4*	
Mean	21.6	22.1	25.9	31.0	(Group 3 compared)

The last extreme groups to be tested are groups 2 and 1. The mean difference, between groups 2 and 1 is not significant. Therefore, a common line is drawn to indicate the finding.

	Gr. 1	*Gr. 2*	*Gr. 3*	*Gr. 4*	
Mean	21.6	22.1	25.9	31.0	(Group 2 compared)

(In the row above, 21.6 and 22.1 are underlined by a common line.)

To summarise the whole results:

1. Group 4 is reliably superior to groups 3, 2 and 1.
2. Group 3 also is reliably superior to groups 2 and 1.
3. Groups 1 and 2 did not show any difference between them.

The results may be presented in a tabular column.

Table 12.5 : Mean difference between groups R_p values and level of significance

Values	*Groups*	*Gr. 1*	*Gr. 2*	*Gr. 3*	*Gr. 4*
Mean diff.	Gr. 1		0.5	4.3	9.4
R_p value		X	2.9	3.05	3.14
Level of sig.			N.S.	0.05	0.05
Mean diff.	Gr. 2			3.8	8.9
R_p value			X	2.91	3.05
Level of sig.				0.05	0.05
Mean diff.	Gr. 3			X	5.1
R_p value					2.01
Level of sig.					0.05
Mean diff.	Gr. 4				
R_p value					X
Level of sig.					

In applying Duncan's Range Test, first we must observe the mean difference. If the mean differences are high, approximately above 5 then we can straight away go for estimating one per cent level of significance provided the R_p values are low. Otherwise, it is better to go for 5 per cent level of significance.

The procedure can be followed for groups 5 or 6 or any number of groups. It is important that the steps described here should be followed carefully. The important values used in Duncan's Range Test are S_e = Error variance. The equation should be used for computing this value. The next important value is the r_p value. This is obtained from the *Duncan's Ratio Table* for degrees of freedom (Table 12.2). $N - r$, N = number of total participants in the experiment and r is the number of groups. The third important value is the R_p. The R_p is called 'the least significant range'. This is obtained by applying the equation $R_p = S_e\, r_p \sqrt{\frac{1}{N}}$. The S_e value is constant. N is the number of participants in a group. That is for every group, one R_p value should be worked out. The SS value is the summary of squares. This is worked out for each group.

The formula for SS is : $SS = \Sigma X^2 - \frac{(\Sigma X)^2}{n}$

The formula for S_e is $\sqrt{\dfrac{SS_1 + SS_2 + SS_3 + SS_4 \ldots S_r}{(n_1 - 1) + (n_2 - 1) + (n_3 - 1) + (n_4 - 1)}}$

The researcher should make himself familiar with all the formula and also the procedures step by step before he makes use of Duncan's Range Test. Otherwise, it is better that he should go for the classical analysis of variance and '*t*' test or critical ratio for comparing the means of multiple groups. At this state the reader might have had some idea about the application of Duncan's Range Test.

Therefore, following the same procedure, let us have one more example illustrated involving more than three groups. In this example, let us have a five group experimental design.

Duncan's Multi-group Design—Five Groups with Unequal N's

The researcher, for example, wanted to break the sample into five groups on the basis of some variable, say, occupation and to know whether there is any difference in the opinion among them regarding the need of higher education for woman. All the participants have answered five questions on a five point Likert type scale, with five response categories such as strongly agree, agree, undecided, disagree and strongly disagree. The five occupational groups are Company Executives, Engineers, Factory Managers, Teachers and Doctors. Their scores are as follows. The number of participants are not equal in these groups. The highest score is 25 most favourable, the lowest score is 5, most unfavourable for higher education for women.

Table 12.6 : Scores of five values of unequal groups

1 C. Executives		*2* Engineers		*3* F. Managers		*4* Teachers		*5* Doctors	
X	X^2	X	X^2	X	X^2	X	X^2	X	X^2
10	100	11	121	14	196	18	324	18	324
14	196	12	144	14	196	14	196	14	196
10	100	10	100	14	196	15	225	14	196
10	100	8	64	12	144	14	196	20	400
13	169	16	256	13	169	19	361	18	324
12	144	15	225	15	225	17	289	18	324
13	169	14	196	13	169	15	225	15	225
		12	144	12	144	13	169	16	256
		13	169	16	256			17	289
				17	289			16	256
								18	324
82 ΣX	978 ΣX^2	111 ΣX	1419 ΣX^2	140 ΣX	1984 ΣX^2	125 ΣX	1985 ΣX^2	184 ΣX	3114 ΣX^2
N 7		9		10		8		11	
X 11.7		12.3		14.0		15.4		16.70	

The next step is to compute the sum of squares.

$$SS_1 = \Sigma X^2 - \frac{(\Sigma X)^2}{n}$$

$$SS_1 = 978 - \frac{(82)^2}{7} = 978 - \frac{6774}{7} = 978 - 960.57 = 17.43$$

$$SS_2 = 1419 - \frac{(111)^2}{10} = 1419 - \frac{12321}{9} = 1519 - 1369 = 50.00$$

$$SS_3 = 1984 - \frac{(140)^2}{10} = 1984 - \frac{19600}{10} = 1984 - 1960 = 24.00$$

$$SS_4 = 1985 - \frac{(125)^2}{8} = 1985 - \frac{15625}{8} = 1985 - 1953.13 = 41.87$$

$$SS_5 = 3114 - \frac{(184)^2}{11} = 3114 - \frac{44856}{11} = 3114 - 3077.82 = 36.18$$

Table 12.7 : Values of r_p for Duncan's Range Test (significance level = 1 per cent)

	Number of Groups														
df	2	3	4	5	6	7	8	9	10	12	14	16	18	20	50	100
1	90.0	90.0	90.0	90.0	90.0	90.0	90.0	90.0	90.0	90.0	90.0	90.0	90.0	90.0	90.0	90.0
2	14.0	14.0	14.0	14.0	14.0	14.0	14.0	14.0	14.0	14.0	14.0	14.0	14.0	14.0	14.0	14.0
3	8.26	8.5	8.6	8.7	8.8	8.9	8.9	9.0	9.0	9.0	9.1	9.2	9.3	9.3	9.3	9.3
4	6.51	6.8	6.9	7.0	7.1	7.1	7.2	7.2	7.3	7.3	7.4	7.4	7.5	7.5	7.5	7.5
5	5.70	5.96	6.11	6.18	6.26	6.33	6.40	6.41	6.5	6.6	6.6	6.7	6.7	6.8	6.8	6.8
6	5.24	5.51	5.65	5.73	5.81	5.88	5.95	6.00	6.0	6.1	6.2	6.2	6.3	6.3	6.3	6.3
7	4.95	5.22	5.37	5.45	5.53	5.61	5.69	5.73	5.8	5.8	5.9	5.9	6.0	6.0	6.0	6.0
8	4.74	5.00	5.14	5.23	5.32	5.40	5.47	5.51	5.5	5.6	5.7	5.7	5.8	5.8	5.8	5.8
9	4.60	4.86	4.99	5.08	5.17	5.25	5.32	5.36	5.4	5.5	5.5	5.6	5.7	5.7	5.7	5.7
10	4.48	4.73	4.88	4.96	5.06	5.13	5.20	5.24	5.28	5.36	5.42	5.48	5.54	5.55	5.55	5.55
11	4.39	4.63	4.77	4.86	4.94	5.01	5.06	5.12	5.15	5.24	5.28	5.34	5.38	5.39	5.39	5.39
12	4.32	4.55	4.68	4.76	4.84	4.92	4.96	5.02	5.07	5.13	5.17	5.22	5.24	5.26	5.26	5.26
13	4.26	4.48	4.62	4.69	4.74	4.84	4.88	4.94	4.98	5.04	5.08	5.13	5.14	5.15	5.15	5.15
14	4.21	4.42	4.55	4.63	4.70	4.78	4.83	4.87	4.91	4.96	5.00	5.04	5.06	5.07	5.07	5.07
15	4.17	4.37	4.50	4.58	4.64	4.72	4.77	4.81	4.84	4.90	4.94	4.97	4.99	5.00	5.00	5.00
16	4.13	4.34	4.45	4.54	4.60	4.67	4.72	4.76	4.79	4.84	4.88	4.91	4.93	4.94	4.94	4.94
17	4.10	4.30	4.41	4.50	4.56	4.63	4.68	4.72	4.75	4.80	4.83	4.86	4.88	4.89	4.89	4.89
18	4.07	4.27	4.38	4.46	4.53	4.59	4.64	4.68	4.71	4.76	4.79	4.82	4.84	4.85	4.85	4.85
19	4.05	4.24	4.35	4.43	4.50	4.56	4.61	4.64	4.67	4.72	4.76	4.79	4.81	4.82	4.82	4.82

(Contd.)

Table 12.7 (*Contd.*)

	Number of Groups															
df	2	3	4	5	6	7	8	9	10	12	14	16	18	20	50	100
20	4.02	4.22	4.33	4.40	4.47	4.53	4.58	4.61	4.65	4.69	4.73	4.76	4.78	4.79	4.79	4.79
22	3.99	4.17	4.28	4.36	4.42	4.48	4.53	4.57	4.60	4.65	4.68	4.71	4.74	4.75	4.75	4.75
24	3.96	4.14	4.24	4.33	4.39	4.44	4.49	4.53	4.57	4.62	4.64	4.67	4.70	4.72	4.74	4.74
26	3.93	4.11	4.21	4.30	4.36	4.41	4.46	4.50	4.53	4.58	4.62	4.65	4.67	4.69	4.73	4.73
28	3.91	4.08	4.18	4.28	4.34	4.39	4.43	4.47	4.51	4.56	4.60	4.62	4.65	4.67	4.72	4.72
30	3.89	4.06	4.16	4.22	4.32	4.36	4.41	4.45	4.48	4.54	4.58	4.61	4.63	4.65	4.71	4.71
40	3.82	3.99	4.10	4.17	4.24	4.30	4.34	4.37	4.41	4.46	4.51	4.54	4.57	4.59	4.69	4.69
60	3.76	3.92	4.03	4.12	4.17	4.23	4.27	4.31	4.34	4.39	4.44	4.47	4.50	4.53	4.66	4.66
100	3.71	3.86	3.98	4.06	4.11	4.17	4.21	4.25	4.29	4.35	4.38	4.42	4.45	4.48	4.64	4.65
∞	3.64	3.80	3.90	3.98	4.04	4.09	4.14	4.17	4.20	4.26	4.31	4.34	4.38	4.41	4.60	4.63

Experimental Psychology, 3/E by McGuigan, ©1978, Adapted by permission of Prentice-Hall, Inc., Upper Saddle River, NJ. p. 255.

The third step is to compute the error variance from the sums of squares.

$$S_e = \sqrt{\frac{17.43 + 50.00 + 24.00 + 31.87 + 36.19}{(7-1)+(9-1)+(10-1)+(8-1)+(11-1)}} = \sqrt{\frac{159.48}{40}}$$

$$= \sqrt{\frac{159.48}{40}} = \sqrt{3.98} = 1.99$$

Next we have to determine the degrees of freedom. It is $(N - r)$ $df = 45 - 5 = 40$.

With the help of the *df* 45 – 5, the r_p values are determined, from the Table 12.2 for groups five, four, three and two. Let us assume 5 per cent level of significance. We go down first in column 5 of Table 12.8 since there are five groups in this test to the figure corresponding to 40 *df*. For five groups the r_p is: 3.17, for four groups the r_p : 3.10, for three groups the r_p : 3.01 and for two groups the r_p is 2.86.

Table 12.7, Value for r_p for Duncan's Range Test for 1 per cent level is applicable only in case, if the r_p values are higher than 3.64.

Table 12.8 : The r_p values for a five group design with 40 df.

Value	*Gr. 1*	*Gr. 2*	*Gr. 3*	*Gr. 4*	*Gr. 5*
r_p	—	2.86	3.01	3.10	3.17
R_p	—	2.03	1.95	2.07	2.09

The procedure up to this point in the Duncan's multi-group design is the same for equal *n*'s and for non-equal *n*'s. The procedure varies little from this point where we compute the R_p values for the five groups with unequal *n*. The equation applied for determining the R_p value for unequal *n* is:

$$R_p = (S_e)\,(r_p)\,\sqrt{\frac{1}{2}\left(\frac{1}{n_1} + \frac{1}{n_2}\right)}$$

Where the S_e = the error variance

r_p = the least significant standardised range between two means

n_1 = Group 1

n_2 = Group 2.

The next step is to order the means from the lowest to the highest for testing the difference between two extreme groups.

Group	*Gr. 1*	*Gr. 2*	*Gr. 3*	*Gr. 4*	*Gr. 5*
Mean	11.7	12.3	14.0	15.4	16.7

Let us now work out the R_p values

For two groups : $R_p = (2.86)(1.99)\sqrt{\frac{1}{2}\left(\frac{1}{7}+\frac{1}{9}\right)} = 5.69\sqrt{0.127} = 2.03$

For three groups : $R_p = (3.01)(1.99)\sqrt{\frac{1}{2}\left(\frac{1}{9}+\frac{1}{10}\right)} = 5.99\sqrt{0.11} = 1.95$

For four groups : $R_p = (3.10)(1.99)\sqrt{\frac{1}{2}\left(\frac{1}{10}+\frac{1}{8}\right)} = 6.17\sqrt{0.113} = 2.07$

For five groups : $R_p = (3.17)(1.99)\sqrt{\frac{1}{2}\left(\frac{1}{8}+\frac{1}{10}\right)} = 6.31\sqrt{0.11} = 2.09$

Then enter these values in Table 12.9 under r_p values.

Now we are testing the difference between extreme means of the five groups 5, 4, 3, 2 and 1. The extreme means 16.7 and 11.7 are for groups 5 and 1. The difference is (16.7 – 11.7) = 5.0. The R_p value for group 5 is 2.09. This is lower than the mean difference (5.0). Therefore, we can conclude that group 5 is reliably different from group 1 at 5 per cent level. The next comparison is between groups 5 and 2. This is for the extreme values of 4 groups. The mean difference between them is (16.7 – 12.3) = 4.40. We read from Table 12.8 that the R_p value for 4 groups is 2.07. The mean difference is higher than the R_p value. Hence group 5 is reliably superior to group 2 and the difference is significant at 5 per cent level. The next comparison is with three extreme means of the groups. The difference between groups 5 and 3 is (16.7 – 14.0) = 2.7. The R_p value is 1.95. The mean difference is higher than the R_p value. Hence, we can conclude that the group 5 is significantly different from group 3 at 5 per cent level. The other extreme group with which group 5 is to be compared is group 4. This test is between two extreme groups and the mean difference between groups 5 and 4 is (16.7 – 15.4) = 1.4. The corresponding R_p value is 2.03 which is higher than the mean. Therefore, there is no difference between groups 5 and 4 at a significant level.

After testing the extreme means of the five groups we shall now test the extreme means of four groups 4, 3, 2 and 1. The mean difference between groups 4 and 1 is (15.4 – 11.7) = 3.7. The corresponding

R_p value is 2.07, which is lower than the mean difference. Hence group 4 is reliably different from group 1, and the difference is significant at 5 per cent level. While testing means of the extreme groups 4 and 2 the mean difference between groups 4 and 2 is (15.4 – 12.3) = 3.1. The R_p value is 1.95 which is smaller than the mean difference. Hence, we can conclude that group 4 is significantly different from group 2 at 5 per cent level. The other extreme groups to be tested are groups 4 and 3. These two extreme groups have a mean difference of (15.4 – 14.0) = 1.4. The respective R_p value is 2.03. The mean difference is lower than the R_p value. This means that there is no significant difference between groups 4 and 3.

The third test is, testing the means of three extreme groups 3, 2 and 1. The mean difference between groups 3 and 1 is (14.0 – 11.7) = 2.3. The corresponding R_p value is 1.95. The mean difference is higher than the R_p value. Therefore, it can be concluded that group 3 is reliably different from group 1 and the difference is significant at 5 per cent level. The other extreme groups are groups 3 and 2. The difference between groups 3 and 2 is (14.0 – 12.3) = 1.7. The relevant R_p value is 2.03. The R_p value is higher than the mean difference. Therefore, we can conclude that group 3 is not reliably different from group 2.

The fourth test is with two extreme groups 2 and 1. The mean difference between groups 2 and 1 is (12.3 – 11.7) = 0.6. The relevant R_p value is 2.03. The mean difference is very negligible and the R_p value is very high. Therefore, we can conclude that group 2 is not different from group 1. A common line can be drawn to indicate the results. The common lines indicate that there is no difference between the groups.

	Gr. 1	*Gr. 2*	*Gr. 3*	*Gr. 4*	*Gr. 5*	
Mean	11.7	12.3	14.0	*15.4*	*16.7*	Testing 5 extreme groups
Mean	11.7	12.3	*14.0*	*15.4*	16.7	Testing 4 extreme groups
Mean	11.7	*12.3*	*14.0*	15.4	16.7	Testing 3 extreme groups
Mean	*11.7*	*12.3*	14.0	15.4	16.7	Testing 2 extreme groups

To summarise the whole results:

1. Group 5 is reliably superior to groups 3, 2 and 1, but no reliable difference with group 4.
2. Group 4 is reliably superior to groups 2, 1 and 5 but not with group 3.

3. Group 3 is significantly different from groups 1, 4 and 5 but not reliably significant with group 2.
4. Group 2 is significantly different from groups 3, 4 and 5 and not with group 1.

The results are presented below in Table 12.9 .

Table 12.9 : Mean difference between groups, R_p value and level of significance

Values	*Groups*	*Group 1*	*Group 2*	*Group 3*	*Group 4*	*Group 5*
Mean diff. R_p value L. Sig.	Group 1	X	0.6 2.03 NS	2.3 1.95 0.05	3.7 2.07 0.05	5.0 2.09 0.05
Mean diff. R_p value L. Sig.	Group 2		X	1.7 2.03 NS	3.1 1.95 0.05	4.4 2.07 0.05
Mean diff. R_p value L. Sig.	Group 3			X	1.4 2.03 NS	2.7 1.95 0.05
Mean diff. R_p value L. Sig.	Group 4				X	1.30 2.03 NS
Mean diff. R_p value L. Sig.	Group 5					X

The procedures described above can be applied to any number of groups. The researcher should be cautious with regard to the estimation and the use of the R_p values and with regard to the selection of the correct r_p values. There are certain values in the Duncan's Range Test that may raise doubts, particularly the selection of R_p values for comparing the mean differences. In such cases the researcher is advised to refer to the model illustrations. In Duncan's Range Test comparison is made between the mean difference of extreme groups and the corresponding R_p value.

The Duncan's Range Test has one small disadvantage. Unlike analysis of variance or other methods in Duncan's Range Test the level of significance is already decided. One will have to assume that the extreme means may vary at 5 per cent level or one per cent level. In other methods the level of significance could be determined only after the *F*-ratio, or *t*-ratio or critical ratios are obtained. It is also possible that certain measures may have 5 per cent level of significance and certain measures one per cent level of significance in the same experiment. This can be easily determined in other methods by seeing the ratios. But in Duncan's Range Test the result will show either it is significant at 5 per cent level or at 1 per cent level depending upon the assumption set up at the beginning of the computation. Therefore, it is better to find out the r_p values for both 5 per cent level and for 1 per cent level and compute the R_p values for both 5 and 1 per cent levels. But it also depends upon the mean differences. Whatever, the mean difference is we can go straightaway for Duncan's Range Test. If the mean differences are higher then we can assume that the differences are significant at one per cent level and compute the R_p values.

The analysis of variance itself does not provide us with all the information for multi-group comparisons. You will have to apply other statistical procedures for comparisons. That is why the analysis of variance and the *F*-test are always followed by *'t'* tests and critical ratios. If there are 10 groups to be compared, analysis of variance will not tell us anything between any two groups. It can tell us as a whole whether there is any difference among the groups or not. After this you will have to work out the *'t'* test or critical ratio to see the level of difference between groups. This becomes cumbersome and the table becomes very lengthy. However, these disadvantages were realized by the researchers, they have followed the analysis of variance and the *F*-test followed by *'t'* test till a statistical procedure specifically designed for multiple group comparisons have become prominent among the psychologists in 1960's.

McGuigan (1978) while discussing in detail the merits and demerits of analysis of variance cites illustrations to show how the analysis of variance and the *F* test could help us only to reject or accept the null hypothesis and nothing more. If we are able to reject the null hypothesis then we come to know that some significant difference is there between groups. This is what the *F* test could tell us. But we do not know where the difference between groups lies. In a multi-group design not all the groups will have difference with other groups. There may be difference between some groups and there may not be any reliable difference between some groups. Both types of results may be present in the analysis of variance and *F*-test. Unless we go for *'t'*-test we cannot say where the differences are falling. When the researchers face such critical junctures Duncan's Range Test comes for rescue.

The researcher's problem is to determine whether his experimental groups differ reliably. One way of comparing the groups is to use mean values. But the researcher wants to know what statistical method or procedure is quite appropriate for this type of situation. McGuigan (1978) states it in a very lamenting manner that it is very unfortunate that there is much disagreement among the psychologists and statisticians with regard to the correct answer to this question raised by the researcher. He minimized the controversies on this issue by suggesting a more appropriate method, 'Duncan's Range Test' for solving this problem. Duncan's Range Test provides all possible chances to compare between pairs of groups. It is less laborious than *F*-test and '*t*'-test in computation. Further it also provides a better and more reasonable probability level in comparisons than the '*t*'-tests.

13

ANALYSIS OF VARIANCE AND THE *F* TEST

Introduction

In psychological measurements researchers quite often use the '*t*'-test for testing the significance of the results of their experiments. If the sample is small, then '*t*'-test is used invariably, otherwise, if the sample is large critical ratio is applied. This is applicable adequately to any experiment that involves comparing two groups and testing of a single mean difference. But if the groups are three or more then the problem arises. The same '*t*'-test and critical ratio cannot be applied to such experimental group means. Suppose, for example, we had designed an experiment involving three groups, group 1, group 2 and group 3 and each group was tested under different experimental conditions. It is still possible to use *t*-test and critical ratio to estimate the difference between the three group means, by comparing groups 1 and 2, 2 and 3, and 1 and 3. If it is like this, then there is no problem. It is simple. But if the groups are 5 or more or 10, then the comparisons become too many and the simple procedure applied in *t*-test or critical ratio needs to be replaced with some other method of testing.

When and Why we go for Analysis of Variance?

How the problem arises when the groups are many, for example 5? The comparisons between groups are $\frac{N(N-1)}{2} = \frac{5(5-1)}{2} = \frac{5 \times 4}{2} = \frac{20}{2} = 10$. Suppose we had 10 groups then the number of comparisons become $\frac{10(10-1)}{2} = \frac{10 \times 9}{2} = \frac{90}{2} = 45$. The *t*-test and critical ratio testing procedures become very laborious and cumbersome, as we are making in this only one to one comparisons. The procedure, that could solve the problem of testing several groups together whether there is any difference among the groups is the analysis of variance and the F-ratio distribution. The analysis of variance is a statistical procedure by which the data could be treated together. Till 1960's before Duncan's Multiple Range Test become popular, analysis of variance has the most preferred testing procedure in multi-group designs. Analysis of variance is not directly dealing with dispersions like standard deviations and standard errors. It mainly deals with means and variances.

What is Analysis of Variance?

Analysis of variance is a measure of dispersion obtained by taking the mean of the squared deviations of the observed values from their mean in a frequency distribution. In analysis of variance the total sum of squares of the sample may be analysed into points i.e. into groups of two, or three or four or any number. It is a single composite test that can test simultaneously the whole sample distribution obtained and conclude the nature of the sample, that is, whether the variances from different samples are identical or homogeneous or heterogeneous. It also tests several samples of the same character and determines if there is any difference between the means of the groups and also whether the sample means vary from the population mean.

In a simple case the *total sum of squares* (1) is treated or analysed into two parts. The sum of squares based upon variation *within groups* (this may be 3 or 4 or 5 or any number of groups), (2) and the sum of squares based upon the variation between group means, (3) From the last two values, the sums of squares, the population variance is estimated. It is based upon the assumption that the groups or samples making up the total measurements are random samples and drawn from a normal population. But the two estimates, (within groups and between groups) of the variance of the population vary within the sample limit. The null hypothesis may be tested on the basis of the difference between the variance. It is also expected that the second one, variation between groups, is larger than the first one, variation within groups. To get the ratio between the two, we will have to divide the larger by the smaller variance. The answer is the *F*-ratio, which may be significant at 5 per cent level or 1 per cent level. This is decided by going through the Table 13.1 figure with degrees of freedom.

If the obtained *F*-ratio value is equal to the Tabled value or goes higher than that then the null hypothesis cannot be defended that the samples are drawn from the same normal population. This means that the samples drawn for the two estimates vary either in means or in variances. Therefore, the null hypothesis is rejected. It can be identified whether the difference is caused by the means or variances. If the variances are varying to some extent then we can conclude that the means are the causes for the difference. If this case occurs further treatment to the data may identify the variances that cause the difference. Therefore, the three important components associated with analysis of variance are, the total sum of squares composed of several groups, and analysing this into two components, variation within the groups and variations between group means. This the basic principle of analysis of variance. Based upon this wc shall illustrate it with some examples.

Experimental Illustration

In an experiment conducted on a small group of students drawn randomly from three schools the following data were obtained in a memory test. The sample consisted of 30 students and there were 10 students in each group. They are all of the same age group but drawn from different localities. The data are the amount of material retained and presented by the students in the memory test.

Analysis of variance can be applied to test the mean differences of any number of randomized groups. In testing the means of several groups the test of analysis of variance is to test the homogeneity of the samples. But the most simple application is to test the means of two randomised groups. In the case of two randomized groups the *t*-test is more appropriate rather than analysis of variance. Therefore, analysis of variance may be applied to more than two randomized groups.

It was already stated that there are three important parts in analysis of variancc and all three parts have separate equation. The first one is computing the total sum of squares (*SS*). The equation is:

$$\text{Total } SS = X_1^2 + X_2^2 + X_3^2 + \ldots + X_r^2 - \frac{(X_1 + X_2 + X_3 + \ldots + X_r)^2}{N}$$

The subscript *r* stands for continuation and adding the values of the sums X^2 and the sum of *X* in the equation for as many groups as we can have in the experiment.

Table 13.1 : Showing the scores of 3 groups of students in the memory test

Gr. 1		*Gr. 2*		*Gr. 3*		
X	X^2	X	X^2	X	X^2	
4	16	6	36	12	144	
5	25	7	49	10	100	
6	36	6	36	10	100	
5	25	5	25	10	100	
6	36	5	25	12	144	
4	16	6	36	12	144	
6	36	6	36	10	100	
4	16	7	49	12	144	
5	25	6	36	12	144	
5	25	6	36	10	100	
$X - 50$	X^2 256	60	X^2 364	110	X^2 1220	$X = 220$
$N - 10$		10		10		$X^2 = 1840$
X 5.0		6.0		11.0		$X = 8.67$
$(X)^2$	2500		3600		12100	$(X)^2 = 18200$

The second step is to break the total *SS* into components. As we have already stated the other two important components (other than the total *SS*) are sum of squares (*SS*) among the groups and Sum of Squares (*SS*) within the groups.

The equation for sum of squares among or between groups is:

$$SS = \frac{(X_1)^2}{n_1} + \frac{(X)^2}{n_2} + \frac{(X)^2}{n_3} + \ldots + \frac{(X_r)^2}{n_r} - \frac{(X_1 + X_2 + X_3 + \ldots + X_r)^2}{N}$$

The other important component is sum of squares within groups. The equation for computing the *SS* within groups is:

Sum of squares (*SS*) within group = Total *SS* – *SS* among or between groups.

There may be several groups in the experimental situation as randomized groups. In such cases the *SS* for several groups should be computed. Hence, the ΣX and ΣX^2 should be computed for each group separately. The values ΣX_1, ΣX_2, etc. are all sums of dependent variable for groups 1, 2 and so on. The values ΣX_1^2, ΣX_2^2, ΣX_3^2 are the sums of the dependent variable for groups 1, 2, 4 and so on. The capital N is the sum total of all the participants in the experiment. The small n refers to the number of participants in a single group.

Now let us apply the equation to our data. Since there are 3 groups the equation required for Total *SS* is:

$$= (\Sigma X_1^2 + \Sigma X_2^2 + \Sigma X_3^2) - \frac{(\Sigma X_1 + \Sigma X_2 + \Sigma X_3)^2}{N}$$

In the computation of Total *SS* we can see that the sum of *X* for group 1 is 40, for group 2 it is 60 and for group 3 it is 80. This we may say as $\Sigma X_1 = 40$, $\Sigma X_2 = 60$ and $\Sigma X_3 = 80$. In the same way for ΣX^2 values may be stated as $\Sigma X_1^2 = 168$, $\Sigma X_2^2 = 364$ and for $\Sigma X_3^2 = 1224$.

$$\text{Total } SS = (256 + 364 + 1220) - \frac{(50 + 60 + 110)^2}{30} = 1840.0 - 1613.3$$

$$= 226.7$$

In the next step we are computing the sum of squares (*SS*) among or between groups. The specific equation for this is:

Among or Between

$$SS = \frac{(\Sigma X_1)^2}{n_1} + \frac{(\Sigma X_2)^2}{n_2} + \frac{(\Sigma X_3)^2}{n_3} - \frac{(\Sigma X_1 + \Sigma X_2 + \Sigma X_3)^2}{N}$$

$$= \frac{(50)^2}{10} + \frac{(60)^2}{10} + \frac{(100)^2}{10} - \frac{(50 + 60 + 110)^2}{30}$$

$$= 250 + 360 + 1210 - 1613.3$$

$$= 1820 - 1613.3 = 206.7$$

The third step is to compute the *SS* for within groups. The equation is:

SS within groups = Total *SS* – *SS* among or between groups.

$$= 226.7 - 206.7 = 20.0$$

It can be noted down here that the sum of squares within the groups plus the sum of squares between the groups is equal to the total sum of squares (Table 13.2).

Table 13.2 : Summary of the computations of analysis of variance

Source of variation	*Sum of Squares*	*df*	*Mean Square*	*F-ratio*	*Level of sig.*
Between groups	206.7	2	103.35	139.7	P 0.001
Within groups	20.0	27	0.74		
Total	226.7	29			

We are now discussing analysis of variance. But we have been using different terms and not variance. The sample values are referred to as mean squares and not as variances. The mean squares within the groups are the estimates of variances of the population. The estimation of mean squares is by dividing the sum of squares by the respective degrees of freedom. There are two mean squares here, one is computed from between groups (103.35) and the other from within groups (0.74). We may state the mean square or the population variance is higher in the case of between groups than the population variance within groups.

The next step is to estimate the degree of relation between the two mean squares or the population variances. How far the variances between groups is greater than the variances within groups? To test this we must use the *F*-test. The *F*-test is a statistical treatment designed by Ronald A. Fisher and named after him with the first letter of his name Fisher (*F* for Fisher).

The equation is:

$$F = \frac{\text{Mean square between or among groups}}{\text{Mean square within groups}}$$

In this equation the numerator shows the differences between or among groups and the denominator shows the variance error of the experiment. Actually the numerator is the effect of the independent variable. If the independent variable is not having any effect upon the dependent variable then the dependent variable will be consisted of only error variance and not true variance. In such case the numerator will be the same size of the denominator and the result will not have any significance. The ratio will be almost 1.0 only.

In our example, the *F*-ratio is found to be 139.7. This means the effect of independent variable is very much and the error variance in denominator is too small. To know at what level this *F*-ratio is significant we must estimate the degrees of freedom. We need three degrees of freedom for this test. The same three steps we must follow here. First we estimated the total *SS* and this was divided into parts as *SS* among or between and *SS* within.

So we have Total $df = N - 1 = 30 - 1 = 29$.

SS among or between groups – the $df = r - 1 = 3 - 1 = 2$

SS within groups – the $df = N - r = 30 - 3 = 27$

The *df* between groups and within groups become the total (2 + 27 = 29). Now let us see at what level our *F*-ratio 139.7 is significant. Let us go for the Fisher and Yates Table V (Appendix C). This table gives all the three levels 0.05 level, 0.1 level and 0.001 level. To find out the *P* value, let us go across the top of the table on the column for 2 degrees of freedom among groups and down to the left side on the rows for 27 degrees of freedom within the groups.

Thus, we find the column number 2 and go down to the row 27. We can find against 2 the row number 27 below the column number 2 the *P* value light faced type 3.35 and 5.49 for 5 and 1 per cent and bold face type value 9.02 for 0.001 level. The computed *F*-ratio 139.7 exceeds the bold face type Table value 9.02 which is significant at 0.001 level. Therefore, the null hypothesis is rejected. We can conclude that the three groups differs reliably at 1 per cent level.

Interpretation

The independent variable exerted much greater influence upon the dependent variable. Since the mean square of between groups is much greater than the mean square of within groups the variances cause greater difference between groups. They show more variation than can be attributed to random sampling from a common population mean. Every group mean differs from the other group means significantly. But what variables cause the difference reliably between groups are not known.

If we want to know this the sample should be broken into variables wise and the statistics called *t* test should be applied. The analysis of variance is applied to see whether the sample as a whole is of homogeneous or heterogeneous group. Since the *F*-ratio is highly significant the sample as whole is heterogeneous in nature. But it is also observed in the values of analysis of variance that there is no reliable difference within the groups. This means that the participants within the groups did not vary very much from others since the mean square within group is very small, less than one. But if the mean square of the within group is greater then the mean square of between groups then we accept the null hypothesis. The error variance is very small in this study. All the three groups vary from each other significantly in the memory test.

An Alternative Procedure

There is another alternative procedure of analysis of variance by which all the three important parts are used to estimate the values. Unlike the conventional method discussed earlier, in this method, apart from total sum of squares of measurements the sum of squared deviations within each group from its mean and the sum of squares between groups are computed and the values are obtained. This form of analysis is suggested by Edwards (1961) and supposed to be the simplest form. There are three specific equations:

The total sum of squares:

$$\Sigma X_1^2 + \Sigma X_2^2 + \Sigma X_3^2 - \frac{(\Sigma X_1 + \Sigma X_2 + \Sigma X_3)^2}{N}$$

With regard to total *SS* there is no difference between the previously quoted equation and this.

The second equation is for sum of squared deviations within each groups:

$$SS = \Sigma X_1^2 - \frac{(\Sigma X_1)^2}{n_1} + \Sigma X_2^2 - \frac{(\Sigma X_2)^2}{n_2} + \Sigma X_3^2 - \frac{(\Sigma X_3)^2}{n_3}$$

This equation was not given in the previous procedure. The *SS* within groups was subtracted from the total *SS* and *SS* between groups. The method may be also called deviation method.

The third equation is sum of squares between groups. The specific equation is evolved by using the mean of each group. The deviation of a group mean $\overline{X}_i$ from the combined group mean $\overline{X}$ is stated as *d*. Hence, the equation is : $d_i^2 = (\overline{X}_i - \overline{X})^2$

Where d_i = deviation of the group mean $\overline{X}_i$ from the combined mean $\overline{X}$. Every squared deviation of the equation given above is based upon this observations or cases. These deviations are multiplied by n_i, the number of cases in each group.

Then the equation is: $n_i d_i^2 = n_i (\overline{X}_1 - \overline{X})^2$

Here the subscript *i* in the equation is referring to the groups 1, 2 and 3. Let us now apply these equations on the same data as given in Table 13.1 earlier and verify whether we get the same conclusions.

Let us first determine the total sum of squares according to this procedure. Towards it we are combining the scores of the three groups and consider them as one set. The specialized formula is:

$$\text{(1) Total } SS = \Sigma X_1^2 - \Sigma X_2^2 + \Sigma X_3^2 - \frac{(\Sigma X_1 + \Sigma X_2 + \Sigma X_3)^2}{N}$$

$$= 256 + 364 + 1220 - \frac{(50 + 60 + 110)^2}{30}$$

$$= 1840 - 1613.33 = 226.7$$

Table 13.3 : Showing the *X* and squares of X^2 of scores on the memory test for 3 groups

Group 1		*Group 2*		*Group 3*	
X	X^2	*X*	X^2	*X*	X^2
4	16	6	36	12	144
5	25	7	49	10	100
6	36	6	36	10	100
5	25	5	25	10	100
6	36	5	25	12	144
4	16	6	36	12	144
6	36	6	36	10	100
4	16	7	49	12	144
5	25	6	36	12	144
5	25	6	36	10	100
Σ 50	256	60	364	110	1220
$\overline{X}$ 5.0		6.0		11.0	

(2) The equation for *SS* within groups:

$$= \Sigma X_1^2 - \frac{(\Sigma X_1)^2}{n_1} + \Sigma X_2^2 - \frac{(\Sigma X_2)^2}{n_2} + \Sigma X_3^2 - \frac{(\Sigma X_3)^2}{n_3}$$

$$= 256 - \frac{(50)^2}{10} + 364 - \frac{(60)^2}{10} + 1220 - \frac{(110)^2}{10}$$

$$= (256 - 250) + (364 - 360) + (1220 - 1210) = 6 + 4 + 10 = \mathit{20}$$

(3) The equation for *SS* between groups is $n_i\, d_i^2 = n_i\ (\overline{X}_i - \overline{X}\ )^2$

$$= n_1\ (\overline{X}_1 - \overline{X}\ )^2 + n_2\ (\overline{X}_2 - \overline{X}\ \ )^2 + n_3\ (\overline{X}_3 - \overline{X}\ \ )^2$$

The mean of three groups : $\frac{50}{10} = 5.0 \quad \frac{60}{10} = 6.0 \quad \frac{110}{10} = 11.0$

The mean of the combined groups $= \frac{5+6+11}{3} = \frac{22}{3} = \mathit{7.33}$

The sum of squares between groups= $n_i\, d_i^2 = 10(5 - 7.33)^2 \quad = 54.29$

$$= n_2\, d_2^2 = 10(6 - 7.33)^2 \quad = 17.69$$

$$= n_3\, d_3^2 = 10(11 - 7.33)^2 \quad = 134.69$$

The sum of three values is 54.29+17. 69+134.69=206.67 or *206.7*

Then let us arrange all the three sums of squares obtained by the alternate formula given above.

Table 13.4 : Summary of the sums of squares of the Analysis of variance

Source of variation	*Sum of Squares*	*df*	*Mean Squares*	*F-ratio*	*Level of sig.*
Between groups	206.7	2	103.35	139.7	P < 0.001
Within groups	20.0	27	0.74		
Total	226.7	29			

In the alternative procedure the equation for sum of squares for between groups and for within groups vary from the first one (Table 13.4). But we get the same values in all three components. In this procedure if the three means were equal to the combined group mean then the sum of squares within groups would be equal to the total sum of squares. If such case occurs then the sum of squares between groups would have zero value.

More than Two Randomised Group Design

Let us illustrate the application of analysis of variance with another example with more than two randomized group design. Students from 5 higher secondary schools have appeared for an Engineering Aptitude Test. There were 75 students on the whole 15 from each school. They were all of the same age group, but drawn from different areas and different socio-economic status groups. They have answered 50 questions and each correct answer carried half mark. The maximum possible marks was only 25. The following were the marks obtained by the 5 groups students (*Table 13.5*).

The same equation stated above may be applied to this data. We first determine the total sum of squares combining the scores of five groups and consider them as one set. The specialized formula is:

$$(1)\ \text{Total } SS = \Sigma X_1^2 - \Sigma X_2^2 + \Sigma X_3^2 + \Sigma X_4^2 + \Sigma X_5^2 - \frac{(\Sigma X_1 + \Sigma X_2 + \Sigma X_3 + \Sigma X_4 + \Sigma X_5)^2}{N}$$

$$= 764 + 2720 + 4620 + 5512 + 7730 - \frac{(102 + 200 + 236 + 286 + 340)^2}{75}$$

$$= 21346 - 18065.28 = 3280.72$$

Table 13.5 : More than two Randomised Group Design

Group 1		*Group 2*		*Group 3*		*Group 4*		*Group 5*	
X	X^2	X	X^2	X	X^2	X	X^2	X	X^2
8	64	9	81	15	225	18	324	24	576
5	25	12	144	14	196	18	324	22	484
5	25	12	144	14	196	18	324	23	529
5	25	12	144	16	256	18	324	22	484
6	36	15	225	15	225	23	529	24	576
5	25	13	169	14	196	18	324	22	484
4	16	12	144	14	196	18	324	19	361
5	25	12	144	15	225	18	324	22	484
6	36	13	169	14	196	19	361	23	529
7	49	15	225	18	324	20	400	24	576
7	49	14	196	17	289	23	529	23	529
8	64	15	225	18	324	18	324	22	484
9	81	14	196	17	229	16	256	23	529
10	100	15	225	18	324	19	361	24	576
12	144	17	289	17	289	22	484	23	529
102	764	200	2720	236	4620	286	5512	340	7730

(2) The equation for *SS* between groups:

$$= \frac{(\Sigma X_1)^2}{n_1} + \frac{(\Sigma X_2)^2}{n_2} + \frac{(\Sigma X_3)^2}{n_3} + \frac{(\Sigma X_4)^2}{n_4} + \frac{(\Sigma X_5)^2}{n_5} - \frac{(\Sigma X_1 + \Sigma X_2 + \Sigma X_3 + \Sigma X_4 + \Sigma X_5)^2}{n}$$

$$= \frac{(102)^2}{15} + \frac{(200)^2}{15} + \frac{(236)^2}{15} + \frac{(286)^2}{15} + \frac{(340)^2}{15} - \frac{(102 + 200 + 236 + 286 + 340)^2}{75}$$

$$= 693.6 + 2666.7 + 3713.06 + 5453.07 + 7706.7$$

$$= 20233.13 - 18065.28 = 2167.85$$

(3) The equation for within groups : Total *SS* – *SS* between groups.

$$= 3280.72 - 2167.85 = 1112.87$$

All the three sum of squares can be put in Table 13.6.

Table 13.6 : Summary of the computation of analysis of variance

Source of variation	*Sum of Squares*	*df*	*Mean Squares*	*F-ratio*	*Level of sig.*
Between groups	2167.85	4	541.96	34.11	P < 0.001
Within groups	1112.87	70	15.89		
Total	3280.72	74			

Let us now find out the degrees of freedom for all three sums of squares. The formula for total *SS* is $N - 1$

The formula for between groups is: $r - 1$

The formula for within groups is: $N - r$

Where N is the total number of participants in the experiment. r refers to the number of groups.

Therefore, we have the *df* for total $75 - 1 = 74$

The *df* for between groups $5 - 1 = 4$

The *df* for within groups $75 - 5 = 70$

The *dfs* between groups and within group is $70 + 4 = 74.$

The *F* ratio is 34.11 which is obtained by dividing the mean squares of between groups (541.96) by the mean square of within groups (15.89). Is the ratio between the mean squares significant? To ascertain this we must use the Fisher and Yates Table V (Appendix C). Whether the *F*-ratio is significant at 5 per cent level or at one per cent level or at 0.001 level can be tested with the Table figure for 4-74 degrees of freedom. The number 4 is the numerator goes from left to right on the table and 70 is the denominator goes from top to down 70. The roman type figure for 4-74 degrees of freedom is 2.50 which is significant at 5 per cent level and the italic type figure is 3.60 which is significant at 1 per cent level. But the bold face type figure is 5.25. The obtained *F*-ratio 34.11 exceeds the Table value 5.25 for 0.001 level. Therefore, we can conclude that the *F*-ratio is reliably significant at 0.001 level. This means that the groups differ among themselves at a very high level. The hypothesis that the samples are random sample drawn from a common normal population is rejected, since groups show more variations than can be attributed to random sampling of the population.

The mean square between groups is higher. This shows that each group differs from the other significantly. But what variance caused the difference between the groups could be determined by *t* test or critical ratio. It is seen that the variance within the group is also high. But we do not know which group is having more dispersion within it. This could be estimated by using the '*t*' or critical ratio. As far as analysis of variance is concerned the important test is whether the five groups are homogeneous or heterogeneous. Are the mean differences due to random sampling or due to variance in the population. Since each group is drawn from varied environmental and socio-economic sources the difference may be due to environmental factor or socio-economic factors.

An Alternative Procedure

The one way classification of analysis of variance can be applied to any number of groups in comparing the means. One example with three groups random design is illustrated in the preceding section. The same data given in Table 13.3 may be used here to illustrate the comparison of five group means in one way classification in an alternative procedure. This procedure seems to be somewhat simpler than the previously described method. This is suggested by Quinn McNemar (1959). The equations also seem to be simple and short. The computation of the sum of squares of deviations is based upon the general principle : $\Sigma X^2 = \frac{1}{N}[N\Sigma X^2 - (\Sigma X)^2]$. The equation for total sum of square is:

$$\Sigma\,\Sigma\,(X - \bar{X}\,)^2 = \frac{1}{N}[N\,\Sigma\,\Sigma X^2 - (\Sigma\,\Sigma X)^2]$$

The equation for within sum of squares is:

$$\overset{k}{\Sigma}\,\Sigma\,(X - \bar{X}_r)^2 = \frac{1}{m}[m\Sigma\,\Sigma X^2 - \Sigma(\Sigma X)^2]$$

The third equation for computing the sum of squares for between groups is: $m\,\overset{k}{\Sigma}\,(\bar{X}_r - \bar{X})^2 = \frac{1}{Km}[K\,\Sigma\,(\Sigma X)^2 - (\Sigma\,\Sigma X)^2]$

There is a technique in this which provides an overall test for the difference between/among several group means considered at the same time. The *'m'* is the formula refers to cases which constitute the groups. They were drawn from a normally distributed population. The subscript *K* refers to the number of groups in the experimental design. It is believed that the results of this technique may not be disrupted by some small departure from normal sample of population or affect the homogeneity of the variances.

The important procedure involved in this is to compute the three sums of squares of deviations. Towards which we need to sum up all the raw scores $\Sigma\Sigma X$, sum up the squares of all the raw scores $\Sigma\Sigma X^2$ and also to sum up the squares of individual group sums, $\Sigma(\Sigma X)^2$. The application of the technique is illustrated with an experimental data obtained from five groups of 75 students, 15 in each group (Table 13.7).

Table 13.7 : Scores of 5 groups on an Aptitude Test (N - 75)

S. No.	*Group 1*		*Group 2*		*Group 3*		*Group 4*		*Group 5*		
	X	X^2	X	X^2	X	X^2	X	X^2	X	X^2	
1	8	64	9	81	15	225	18	324	24	576	
2	5	25	12	144	14	196	18	324	22	484	
3	5	25	12	144	14	196	18	324	23	529	
4	5	25	12	144	16	256	18	324	22	484	
5	6	36	15	225	15	225	23	529	24	576	
6	5	25	13	169	14	196	18	324	22	484	
7	4	16	12	144	14	196	18	324	19	361	
8	5	25	12	144	15	225	18	324	22	484	
9	6	36	13	169	14	196	19	361	23	529	
10	7	49	15	225	18	324	20	400	24	576	
11	7	49	14	196	17	289	23	529	23	529	
12	8	64	15	225	18	324	18	324	22	484	
13	9	81	14	196	17	289	16	256	23	529	
14	10	100	15	225	18	324	19	361	24	576	
15	12	144	17	289	17	289	22	484	23	529	
m	15		15		15		15		15	—	
ΣX	102		200		236		286		340		$\Sigma\Sigma X = 1164$
ΣX^2		764		2720		4680		5512		7730	$\Sigma\Sigma X^2 = 21346$
				40000		55696		81796	115600		$\Sigma(\Sigma X)^2 = 303496$
Mean	6.8		13.3		15.7		19.06		22.67		$\bar{X} = 15.51$

(1) Now, let us compute the total sum of squares by using the formula.

$$\Sigma\Sigma(X - \bar{X})^2 = \frac{1}{N}[N\ \Sigma\Sigma X^2 - (\Sigma\Sigma X)^2]$$

$$= \frac{1}{75}[75(21346) - (1164)^2]$$

$$= 21346 - 18065.28 = 3280.72$$

(2) The formula to compute the sum of squares within groups is:

$$\Sigma\Sigma\,(X-\overline{X}_r)^2=\frac{1}{m}\,[m\,\Sigma\Sigma X^2-\Sigma(\Sigma X)^2]$$

$$=\frac{1}{15}\,[15(21346)-303496]$$

$$=21346-\frac{303486}{15}=21346-20233.07=\mathit{112.93}$$

The formula for computing the sum of squares for between groups is:

$$m\,\Sigma\,(\overline{X}_r-\overline{X})^2=\frac{1}{Km}\,[K\Sigma(\Sigma X)^2-(\Sigma\Sigma X)^2]$$

$$=\frac{1}{(5\times 15)}\,[5(303496)-(1164)^2]$$

$$=\frac{1}{75}\left[5(303496)-\frac{(1164)^2}{75}\right]$$

$$=20233.07-18065.28\quad=\mathit{2167.79}$$

Table 13.8 : Summary of the computation of analysis of variance by deviation method

Source of variation	*Sum of squares*	*df*	*Mean squares*	*F-ratio*	*Level of sig.*
Between or among groups	2167.79	4	541.95	34.11	< 0.001
Within groups	1112.93	70	15.89		
Total	3280.72				

The alternative procedure involves the deviation of scores. The raw scores are summed up for all the subjects in the group and also for all the groups. The sums are summed up for the raw scores and also for the squared scores. The deviations are obtained by using the respective formulas. These are the sums of squares with degrees of freedom. Then the mean squares or the variances are estimated and given in the Table 13.8. The sum of squares for between or among the groups and the sum of squares for within groups are added to give the total sum of squares. The mean squares for among groups 541.95 is divided by the mean squares for within groups. We get the *F*-ratio 34.11 which is reliably significant at 0.001 level. In the alternative method also we get the same result. From the Fisher and Yates Table V (Appendix C) we find that the *F*-ratio 34.11 is significant at 0.001 level.

The mean squares or the variance between or among groups is found to be reliably significant and higher than the within groups variance. This shows that the means of five groups are not drawn from the same population. All the five groups are heterogeneous in characteristics. Therefore, the null hypothesis is rejected. The sample is not drawn randomly from the same simple population. The variances between or among groups are higher than the variances within groups. This means that the variation between individuals in the group is smaller, whereas, the variation between groups is higher.

The analysis of variance and the *F*-test have determined that all the five groups are different from each at 0.001 level. But we do not know to what extent one group varies from the other group. The variance method we are now discussing is having one advantage over other method of analysis variance. If we want to know the difference between one group and another we must go for *t*-test or *CR* techniques. According to these techniques if we compute the *t* or *CR* for any two means then the comparison are $\frac{n(n-1)}{2} = \frac{5 \times 4}{2} = 10$. It is a cumbersome task. Therefore, the variance method suggests less computational procedure. The variance method also suggests that the *t* or *CR* also can be an overall test of significance if the *t* or *CR* is large. Just like the *F*-ratio can provide an overall significance in a multi-group design so also the *t* or *CR* also provide an overall significance between any two groups. The values required for this purpose are also given in the estimation of sum of squares. The important values required are the means of various groups, the mean squares or variances for within groups and the *n*. The formula for testing the mean differences is:

$$t \text{ or } CR = \frac{M_1 - M_2}{\sqrt{\frac{\sigma_1^2}{N_1} + \frac{\sigma_2^2}{N_2}}} = \frac{13.3 - 6.8}{\sqrt{\frac{15.90}{15} + \frac{15.89}{15}}}$$

$$= \frac{6.5}{\sqrt{1.06 + 1.06}} = \frac{6.5}{1.45} = 4.48$$

Where M_1 and M_2 the means of the first two groups σ_1 and σ_2 the standard deviations or variance of group 1 and 2. n_1 and n_2 the number of cases in each group.

The variance (mean squares) estimate is based on the whole sample and hence the, degrees of freedom is 75. Though we compared the groups in the estimation of *CR*, each group having the same *n* as 15, the standard deviation becomes a common factor for all the five groups in the deviation method. Therefore, the *CR* 4.48 for 75 degrees of freedom is significant at one per cent level. Since the *CR* is very high (4.48), we can also make use of the Fisher and Yates Table IV and we can see that the CR is significant at 0.001 level (Appendix B).

The mean difference between groups 2 and 3 is very small and the difference is not significant at any level. But in the remaining 9 comparisons the mean scores differences between groups are high and reliably significant. This is one great advantage of the deviation method. We can compute the *F*-ratio and also the *t*-ratio and critical ratio with the same values.

Factorial Design: Analysis of Variance with Two Independent Variables

We have been discussing so far one way classification of analysis of variance. Though there were 3 groups and 5 groups in the experiment there was only one independent variable which was having influence upon the dependent variable. Even there were more than one variable acting upon the dependent variable we were not giving any importance for them. Samples drawn from the same population vary due to certain independent variables. The independent variable may be varied in two ways. In this case two groups designs may be used. It is called two way classification of problem. It is also possible to vary independent variable in more than two ways. In such cases the multi-groups designs may be used. However, it is also possible to use more than one independent variable in a single experiment. When we have three or more basis for classifying the observations we have factorial analysis of variance designs. These designs are not limited to any classification of analysis. In a single experimental design three or more independent variables may be operating and it is also possible to attribute the effects to those independent variables and the dependent variables may be classified on these basis and analysed. In other words we have discussed earlier that the analysis of variance technique was applied to relatively simple situations. These situations involve the testing of the significance of the overall variation of the means for several groups. The groups were made to differ on the basis of a single variable. These types are referred to a single variable experiments, or one way classification of the problem. This means that groups are differing in one known respect and the groups are compared on a dependent variable. But psychologists also interested in designing experiments in which one set of data should serve for testing the influence of two or more independent variables.

This type of experimental designs have been done for a considerable amount of time. First, to classify the total group in one way, then in another way and then in a third way. Suppose, for example, we are determining the achievement levels of a group of students and we may classify them on the basis of environmental factors such as rural, urban and semi-urban. Secondly, we go for the socio-economic status of the parents and classify them on the basis of parents' educational qualification and income. Thirdly, we may classify the students on the basis of age, and fourthly on the basis of sex and so on. In this procedure one variable is considered at a time and this leads to single variable set up though the same data answered to the effects of different independent variables. In this study of achievement there are factors influencing achievement. Now we could make a double classification by dividing the cases on two of the variables simultaneously or by a triple classification by dividing the cases on three of the variables. At present let us consider a double classification or a two way classification of analysis of variance. It is also called factorial design. We study in the factorial design the interaction of two or more variables simultaneously each varying in several ways.

An Experimental Illustration

Three psychomotor tests of similar type were conducted on 45 students who appeared for Engineering entrance test. Three types of abilities were assessed from them such as Comprehension, Application and Evaluation. Therefore, there were 9 experimental conditions. The students were assigned to one of the 9 experimental conditions at random. Thus, there were 5 students in each group. There were two independent variables in this test. Variable A indicates type of ability which is varied in three ways. The other independent variable is B which indicates the Psychomotor tests. This also is varied in three ways by using three different tests. Thus, in the cell entry we have A_1, A_2 A_3 and in the B variable B_1, B_2 and B_3. We have designated Variable A as type of ability and variable B as the psychomotor tests. Therefore, the first cell entry becomes A_1B_1. This reveals the ability, comprehension in test 1 and A_1B_2 reveals comprehension in test 2 and A_1B_3, comprehension in test 3. The total in row A_1 shows the revealed ability of comprehension by all 3 tests. In the same way each cell is filled with the scores of the independent variable *A* and *B* by varying each variable in three ways.

Table 13.9 : Experimental design with two independent variables studying the outcome of three different tests and the abilities revealed

Variable-A Type of Ability	*Variable-B Psychomotor test*			
	Test 1	Test 2	*Test 3*	
1. Comprehension	A_1B_1	A_1B_2	A_1B_3	A_1
2. Application	A_2B_1	A_2B_2	A_2B_3	A_2
3. Evaluation	A_3B_1	A_3B_2	A_3B_3	A_3

Thus, we have 9 groups with 5 cases in each group. There are 3 comparisons to make with Type of ability and 3 comparisons to make with type of tests. By comparing every experimental condition with every cell in the Table we would have $\frac{n(n-1)}{2} = \frac{9 \times 8}{2} = 36$ comparisons. Among these 36 comparisons we do not know which of the mean differences would be reliably significant. Therefore, we make overall comparisons by the *F* test and then we can go for *t* test for the cases which are having significant differences.

The scores were obtained and treated in a two way classification of analysis of variance and presented in Table 13.10.

The Table 13.10 is completed with scores obtained in 3 tests in three abilities. Now the next step is to calculate the sums of squares as we did in the earlier cases.

Total sum of square is computed by squaring and adding each raw score of all the subjects (*N*). Here the *N* is 45. The raw scores for all the cases are added and squared and subtracted from the total *N*'s scores. The formula is the same:

$$\text{Total sum of squares} = (18)^2 + (14)^2 + (15)^2 + (13)^2 + (15)^2\ (7)^2 + (6)^2$$
$$+ (9)^2 + (10)^2 + (13)^2 + (4)^2 + (4)^2 + (5)^2 + (9)^2 + (8)^2 + (5)^2 + (7)^2$$
$$+ (8)^2 + (10)^2 + (10)^2 + (13)^2 + (13)^2 + (14)^2 + (14)^2 + (16)^2 + (5)^2$$
$$+ (7)^2 + (8)^2 + (9)^2 + (11)^2 + (3)^2 + (3)^2 + (4)^2 + (8)^2$$
$$+ (7)^2 + (6)^2 + (5)^2 + (8)^2 + (9)^2 + (12)^2 + (10)^2 + (12)^2$$
$$+ (12)^2 + (13)^2 + (13)^2 - \frac{(425)^2}{45} = 4663 - 4013.88 = \mathit{649.12}$$

Table 13.10 : Scores of 45 students earned on three psychomotor tests inthree types of abilities

Variable-A Type of Ability	*Participants*	*Variable-B Psychomotor Tests*			*Sum and mean of Ability*
		Test 1	*Test 2*	*Test 3*	
Comprehension	1	18	5	3	
	2	14	7	3	
	3	15	8	4	
	4	13	10	8	
	5	15	10	7	
		75	40	25	140
	Mean	15	8	5	9.33
Application	1	7	13	6	
	2	6	13	5	
	3	9	14	8	
	4	10	14	9	
	5	13	16	12	
		45	70	40	155
	Mean	9	14	8	10.33
Evaluation	1	4	5	10	
	2	4	7	12	
	3	5	8	12	
	4	9	9	13	
	5	8	11	13	
		30	40	60	130
	Mean	6	8	12	8.67
		150	150	125	425
		10.00	10.00	8.33	9.44

Sum of squares between groups

$$= \frac{(75)^2}{5} + \frac{(45)^2}{5} + \frac{(30)^2}{5} + \frac{(40)^2}{5} + \frac{(70)^2}{5}$$

$$+ \frac{(40)^2}{5} + \frac{(25)^2}{5} + \frac{(40)^2}{5} + \frac{(60)^2}{5} + \frac{(385)^2}{45}$$

$$= 1125 + 405 + 180 + 320 + 980 + 320 + 125 + 320$$

$$+ 720 - 4013.88$$

$$= 4495 - 4013.88 = 481.12$$

Sum of squares within groups = 649.12 – 481.12 = 168.00

The values of sum of squares are recorded in Table 13.11 for estimating the *F*-ratio.

Table 13.11 : Analysis of variance of scores on three different tests of three abilities for the groups

Source of variation	*Sum of squares*	*df*	*Mean squares*	*F-ratio*	*Level of sig.*
Between groups	481.12	8	60.14	12.88	< 0.001
Within groups	168.00	36	4.67		
Total	649.12	44			

Now the degrees of freedom is to be estimated.

The formula is: 1. Between groups $= K - 1 = 9 - 1 = 8$

2. Within groups $= n - K = 45 - 9 = 36$

3. Total $= n - 1 = 45 - 1 = 44$

Where K = number of groups

n = number of cases.

We must go through the column of the *F*-table with 8 degrees of freedom/or the numerator and the row entry corresponding to 36 degrees of freedom of the denominator we find the figure 3.04 is the *F* which is significant at 0.01 level. But the obtained *F* is 12.88 which is higher than the table value. Therefore, the *F*-ratio 12.88 is reliably significant at 0.01 level. But if we go through Table V of Fisher and Yates (Appendix-C) which provides the significance level up to 0.001 level, our obtained *F* value 12.88 for 8 and 36 degrees of freedom is significant at 0.001 level. The Fisher and Yates Table V does not provide the numbers serially in denominator. It has an interval of 10 between two denominators. Therefore, we will have to work it out for 8-36, taking the values between 30 and 40. For 8 and 40 it is 4.21 and for 8 and 30 it is 4.58. The formula is $4.58 - 4.21\left(\frac{6}{10}\right) = 0.22 = 4.58 - 0.22 = 4.36.$

The *F*-ratio is 12.88. Therefore, it is significant at 0.001 level. The null hypothesis, therefore, cannot be defended since the groups differ among themselves at 0.001 level. But if there were no differences among the groups or populations, the deviation may occur only in less than one per cent of the time as a result of sampling variation. Therefore, the observed differences are not due to chance between groups.

Test of Significance

In the illustrative experiment we have explained so far is a case of two independent variables varied in three different ways. The results show that the nine groups differ among themselves at 0.001 level. But this information is not sufficient for us. We need more information about the differences in the type of abilities and also the effects of the tests. Which test reveals total abilities? Or which test is more efficient in revealing a particular ability? We do not know all these. Therefore, we must analyse the sum of squares between groups to see whether we can get some more information on this experiment in which the sum of squares for type of abilities should be computed. It is done by squaring the sum of scores for each type of ability, A_1, A_2 and A_3 and dividing each by number of cases and summed up and from which the summed up scores of three abilities are squared and divided by the total number of cases and subtracted. This becomes the sum of squares for the type of abilities.

$$\text{Type of ability} = \frac{(140)^2}{15} + \frac{(155)^2}{15} + \frac{(130)^2}{15} - \frac{(425)^2}{45}$$

$$= 1306.67 + 1601.67 + 1126.67 - 4013.89$$

$$= 4035.01 - 4013.89 = \mathit{21.12}$$

The sum of squares for each test of ability, B_1, B_2 and B_3 are divided each by number of cases. The computation is made in the same way as for abilities.

$$= \frac{(150)^2}{15} + \frac{(150)^2}{15} + \frac{(125)^2}{15} - \frac{(425)^2}{45}$$

$$= 1500 + 1500 + 1041.67 - 4013.89$$

$$= 4041.67 - 4013.89 = \mathit{27.78}$$

The sum of sums of squares for type of abilities and tests is equal to = 21.12 + 27.78 = 48.90

The sum of squares between groups which we have already found is: 481.12

Next step is to estimate the sum of squares of interaction. This can be obtained by subtracting the sum of squares for type of abilities and sum of squares for the tests (Table 13.12). Thus we have

Between groups – (sum of squares for abilities + sum of squares for tests) = Interaction

= 481.12 — 48.90 = 432.22

Let us now explain the results as given in Table 13.12. First of all the total sum of squares are analysed into two parts. One part is dealing with variation between each of the cells or groups (Table 13.10). The other part is dealing with variation within each of the groups. The next step is to analyse sum of squares between groups. One part can be referred to variation between psychomotor tests and another part can be referred to variation between types of ability. The third part is called interaction between variables *A* and *B* or tests and types of ability as it becomes the results of the joint result of a particular test and a particular types of ability. Table 13.12 shows the total sum of squares and the number of degrees of freedom and what is done with them. There are 9 experimental groups when two independent variables are varied in three different ways.

Table 13.12 : Analysis of variance of scores on measures of type of abilities for groups tested by three different psychomotor tests

Source of variation	*Sum of squares*	*df*	*Mean squares*	*F-ratio*	*Level of sig.*
Type of ability	21.12	2	10.56	2.26	NS
Psychomotor tests	27.78	2	13.89	2.97	NS
Interaction	432.22	4	108.06	23.14	0.001
Between groups	481.12	8	60.14		
				12.88	0.001
Within groups	168.00	36	4.67		
Total	649.12	44			

There are 5 subjects in each group, making the total subjects 45. Thus we have 4 degrees of freedom within each of these groups and for 9 groups the degrees of freedom becomes $9(5 - 1) = 9 \times 4 = 36$. In the preceding section, in analysis of variance since there had been 9 experimental groups the degrees of freedom had been $(K - 1) = (9 - 1) = 8$. The degrees of freedom between groups is 8 and within groups is $(n - K) = 45 - 9 = 36$, which make up the total of $(n - 1) = 44$ $(36 + 8)$.

In the summary of results as given in Table 13.12, further analysis is made, that is, the sum of squares between groups on the basis of types of ability and sum of squares for tests and a sum of squares for interaction or residual. The degrees of freedom 8 is to be divided among these sum of squares. In this experimental procedure each sum of squares of the tests and each types of ability are based upon 3 groups. Therefore, each sum of these squares is to have $(K - 1) = 3 - 1 = 2$ degrees of freedom. Among the 8 degrees of freedom 2 degrees of freedom are allotted to tests and 2 allotted to types of ability and the remaining 4 degrees of freedom are allotted to the sum of squares of interaction. Allotment of degrees of freedom:

1. Type of ability = $A - 1 = 3 - 1 = 2$ (A = Type of ability)
2. Psychomotor test = $B - 1 = 3 - 1 = 2$ (B = Test)

3. Interaction = $(A - 1)(B - 1) = 2 \times 2 = 4$
4. Between groups = $K - 1 = 9 - 1 = 8$ (K = total number of exp. groups)
5. Within groups = $n - K = 45 - 9 = 36$ (n = total number of subjects)
6. Total = $n - 1 = 45 - 1 = 44$.

The degrees of freedom for interaction is also obtained by multiplying the number of degrees of freedom allotted to the tests by the number of degrees of freedom allotted to types of ability. Thus, we may have: $(A - 1)(B - 1) = (3 - 1)(3 - 1) = 2 \times 2 = 4$. In Table 13.12, the summary of analysis of variance of scores on three different tests is given. It can be seen that each sum of squares of the variations is divided by the corresponding degrees of freedom for getting the mean squares. The F-ratios for all the three variations such as the type of ability, psychomotor test and interaction are obtained by dividing the respective mean squares by the mean square within groups. The values of F are to be determined on the basis of their number of degrees of freedom. For example, for type of ability the degrees of freedom are 2 and 36. The degrees of freedom for tests also the same, 2 and 36. The degrees of freedom for interaction are 4 and 36. For the degrees of freedom 2 and 36 the table F value is 3.26 at 5 per cent level and 5.25 at one per cent level. But our observed values of F for Type of ability and tests administered are 2.26 and 2.97 respectively. The F-ratio for both the variables are neither significant at one per cent level nor at 5 per cent level. This means that the difference of scores in total ability of groups as tested by the three different tests are not significant. The tests did not reveal any difference in the total ability of the groups. Similarly since the F-ratio for tests is not significant, it may be concluded that the subjects did not perform better in any of the tests.

The F-ratio for interaction based on 4 and 36 degrees of freedom from the F table is 2.63 at 5 per cent and 3.89 at one per cent level. But our observed F-value is 23.14 which is significant at 0.001 level according to Fisher and Yates Table V (Appendix C). Therefore, the F value is highly significant. The meaning of this significance is that the effect of one particular test is more effective with one type of ability than another. We do not know which particular test is more effective upon which type of ability. Therefore, here, again the analysis of variance and the F-test did not provide us with correct information with reference to the particular test and the type of ability. But with the available information we can examine the means of various groups as given in Table 13.10 and decide where the test is efficient. We can see that the test one seems to be highly related to the ability of comprehension, the test two is highly related to the ability of application and the test three is highly related to evaluation.

For further treatment of the data we must go for only the *t*-test. It is true that we are interested in various comparisons of the mean scores of the groups. For which we need the standard errors of the means of the groups and the standard errors of the difference between two means. The formula and illustrations for estimating these two statistics, standard error of a single mean and the standard error of the difference between two means are given elsewhere in this book.

But instead of going for *t*-test it is better to go for Duncan's Range Test which is more convenient method for multi-group experimental designs. McGuigan (1978) states that Duncan's Range Test has several advantages. It can reduce the computational labour and it will estimate the significance of the means even if they are smaller. Therefore, researchers are advised to make use of Duncan's Range Test in multi-group experimental designs, particularly for more than two groups. Experimental illustrations for Duncan's Range Test are given in simplified manner elsewhere in this book.

14

SIGNIFICANCE OF MEAN DIFFERENCES: CRITICAL RATIO (Z-RATIO)

Critical Ratio

The sampling error or standard error of the mean has some other important use in psychological measurements. For example, two different samples drawn from the same population. The probability of these two samples can be determined. We assume that these two samples have the probability of being equal in all respects. Suppose, for example we want to know whether students come from urban areas are better achievers in school performances than the students who come from rural areas. We test 200 urban students and 200 rural students and compute the mean scores for the groups. The two groups have shown different mean scores. Is this difference due to chance or real difference based upon the variables? The difference between the means of samples distribute normally. Statisticians believe that this probability of difference between the means can be tested with the statistics of *critical ratio* and can be stated that the difference is a true difference and not due to chance. In applying the statistics of critical ratio the size of the sample becomes an important factor. If the sample is with larger number of cases in both samples, then the difference between the sample scores may be true difference between the populations from which the samples are drawn. The application of *critical ratio* to tests the probability of difference between two samples may also yield for test of significance. It is possible for us to say the level of significance of the difference. The measures required for the estimation of the critical ratio are, the means of the two samples, the standard deviations of the two samples, and the standard error of the two samples and the standard error of the mean difference. The step by step procedure for estimating the critical ratio is:

$$SEM = \frac{SD}{\sqrt{N-1}} \qquad \ldots (1)$$

$$\sigma\, dM \text{ or } SED = \sqrt{SEM_1^2 + SEM_2^2} \qquad \ldots (2)$$

$$CR = \frac{M_1 - M_2}{\sqrt{SEM_1^2 + SEM_2^2}} \text{ or } \frac{M_1 - M_2}{\sigma dM} \qquad \ldots (3)$$

Where M_1 and M_2 = Mean score of the first sample and the second sample.

SEM_1 = Standard error of the first mean

SEM_2 = Standard error of the second mean.

SD = Standard deviation of a sample mean

N = Number of cases in the sample.

σdM = Standard error of the mean difference.

In accordance with a null hypothesis we assume that the two sampling distributions are equal and there is no difference between the two mean scores, that is, $M_1 - M_2 = \sigma$. The standard error of each mean (*SEM*) is estimated by using the first equation $SD/\sqrt{N-1}$. Then the standard error of the mean difference σdM is computed by using the standard errors of the two means, (Equation 2). The third equation is to estimate the critical ratio $M_1 - M_2/\sigma dM$ or *SED*. The two means are supposed to be from independent samples at random from two different unrelated groups.

If the difference between the two means is established then the null hypothesis is rejected. This means $M_1 - M_2 = \sigma$ is not correct. If difference between the two is established it should be also stated at what level, whether it is 0.05 level or 0.01 level?

An Illustrative Example

In a test conducted upon 200 rural students and 200 urban students in the arithmetic test the following are the statistics. (Table 14.1)

The two groups of students were given the same test and the results are summarised in Table 14.1. The urban students mean is 86.50 with an *SD* of 30.60 and that of rural students mean is 74.60 with an SD of 31.50. The urban students scored 11.19 higher than that of the rural. This difference appears to be significant. The sample differences between means are normally distributed. Now we are concerned about determining whether there is any difference between the two groups and also the approximate size. We have information in Table 14.1, that the difference in the obtained sample is 11.19, disregarding the algebraic sign. The standard error of mean difference is 2.95. Thus, we have $\frac{11.19}{2.95} = 4.03$. This value states that how much standard error of the mean difference, the obtained difference extends from the mean of the distribution. The value 4.03 is stated as the critical ratio or z ratio. The obtained critical ratio or z ratio 4.03 is greater than the table value for two-tail test 3.29 for the degrees of freedom 2 and 199. Therefore, we reject the null hypothesis and we say that the obtained difference is significant at 0.001 level. The conclusion clearly demonstrated the urban-rural difference in arithmetic test.

Table 14.1 : Means and other statistics in the comparison of urban and rural students in arithmetic test

Statistics	*Urban students*	*Rural students*
N	200	200
Mean score X	86.50	74.60
SD or S	30.60	31.50
SEM	2.17	2.23
SEM Diff or σdM	2.95	
$(M_1 - M_2)\ D\overline{X}$	11.19	
Critical Ratio or Z	4.03	

Significance of Mean Differences in Small Samples : *t*-test

In applying statistics to the experimental data we do not apply the same statistics to large samples and to small samples. It has been observed that in the case of small samples, the sampling distributions depart from normality in certain ways. For example, the statistics like distribution of correlation coefficient, proportions and standard deviations need to be slanted. Sometimes the distributions become either leptokurtosis or mesokurtosis. Therefore, in small sample tests the student 't' statistics is applied. The 't' may be defined in the same manner as the z or critical ratio was defined. The 't' is actually a ratio between the difference of two sample means and the standard error of the mean difference. There is normal distribution in the statistic z or critical ratio and the samples in this statistic are always large. But the statistic 't' is applied to small samples and sometimes disregarding the sample size.

In the case of student statistic the distribution becomes leptokurtosis as the number of degrees of freedom decreased due to small sample. As the number of cases or degrees of freedom increase the distribution becomes normal. But there is no clear cut distinction between the large sample and small sample in statistics. There is often one extending over the other. There appears to be some general agreement with regard to the size of the small sample. Some say that any number (*N*) less than 30 is a small sample and some put it on a range of 20 to 30. As the *N* in the small sample decreases below 30 there is some change in the treatment of the data and the result becomes critical.

There is not much difference between the statistical formula applied to *z* or critical ratio and to '*t*' ratio. The '*t*'-test is also computed from the difference between the means. The equation for computing the '*t*' test is:

$$t = \frac{\overline{X}_1 - \overline{X}_2}{\sqrt{\dfrac{SS_1 + SS_2}{(n_1-1)+(n_2-1)}\left(\dfrac{1}{n_1}+\dfrac{1}{n_2}\right)}}$$

The '*t*' test can be also computed by another equation. It is

$$t = \frac{\overline{X}_1 - \overline{X}_2}{\sqrt{\dfrac{\Sigma X_1^2 - \dfrac{(\Sigma X)^2}{n_1} + \Sigma X_2^2 - \dfrac{(\Sigma X)^2}{n_2}}{(n_2-1)+(n_2-1)}\left(\dfrac{1}{n_1}+\dfrac{1}{n_2}\right)}}$$

An Illustrative Example

The estimation of '*t*' can be illustrated with an experimental data. The following are the scores obtained from two groups of students in an achievement test (Table 14.2).

Table 14.2 : Students in achievement test

Respondent Number	*Group 1*		*Respondent Number*	*Group 2*	
	X_1	X_1^2		X_2	X_2^2
1	44	1936	1	35	1225
2	45	2025	2	38	1444
3	50	2500	3	40	1600
4	55	3025	4	44	1936
5	60	3600	5	48	2304
6	66	4356	6	50	2500
7	57	3249	7	54	2916
8	49	2401	8	48	2304
9	56	3136	9	56	3136
10	68	4624			
ΣX_1	550 ΣX_1^2	30852		ΣX_2 413	ΣX_2^2 19365

$$\bar{X}_1 = \frac{550}{10} = 55.0$$

$$\bar{X}_2 = \frac{413}{9} = 45.88$$

$$SS_1 = 30852 - \frac{(550)^2}{10} = 30852 - 30250 = 602$$

$$SS_2 = 19365 - \frac{(413)^2}{9} = 19365 - 18952.1 = 19365 - 18952 = 413$$

$$t = \frac{50 - 45.88}{\sqrt{\frac{602 + 413}{(10-1)+(9-1)}\left(\frac{1}{10}+\frac{1}{9}\right)}} = \frac{9.12}{\sqrt{\left(\frac{1015}{17}\right)\left(\frac{9+8}{90}\right)}}$$

$$= \frac{9.12}{\sqrt{(59.70)\left(\frac{19}{90}\right)}}$$

$$= \frac{9.12}{\sqrt{(59.70)(0.211)}} = \frac{9.12}{\sqrt{12.59}} = \frac{0.12}{3.54} = 2.576 \text{ or } 2.58$$

The degrees of freedom for the '*t*' test may be determined by adding the number of participants in the experiment of the groups one and two. It is 10 + 9 = 19. $df = N - 2$.

Therefore, $N = n_1 + n_2 = 10 + 9 = 19$

$df = 19 - 2 = 17$

Now we must see whether the '*t*' is significant. Turning to the table, for the *t* value 2.58 with 17 degrees of freedom the value close to 2.58 on the column is 2.567 below the *P* value 0.02. The obtained *t* (2.58) is larger than 2.228. From this observation we can conclude that the table value of *t* 2.567 has a *P* of 0.02. Any *t* larger than 2.228 must have smaller *P* of 0.02. Therefore, we can interpret the fact that a *t* of 2.58 has a *P* of less than 0.02 ($P < 0.02$). The results indicate that a difference between the means of the two groups has a probability of less than 0.02. This means that if the experiment has been conducted 100 times then we can expect a difference of this size to occur only two times by chance. We will get the same result 98 times.

15

ESTIMATION OF ERRORS IN MEASUREMENTS

Occurrence of Error in Measurements

In advanced psychological studies the application of inferential statistics become very much essential for various purposes. Inferential statistics are also called population statistics. These statistics are used for drawing generalization about the characteristics of populations through samples. Various sampling techniques have been described in the preceding section. A sample is a smaller representation of a larger population. Since a smaller amount is sufficient to analyse the data, large quantities of material is a waste of time and energy. The two broad types of sampling plans, the probability sampling plans and the non-probability sampling plans, however, carefully planned to be randomized representative, proportionated and stratified are not absolutely free from the error variance. While discussing each method of both the plans mentions were made then and there about the way in which errors are occurring in the sampling.

All the statistical procedures discussed in the earlier sections have had a definite purpose of describing the characteristics of the data in a simple and meaningful way. The values of descriptive statistics need not be under estimated in this respect as the masses of data cannot be made intelligible without the assistance of the means and the standard deviations. But, however, we make use of the values of these descriptive statistics they do not serve the purpose of inferences to be drawn that are essential for the better interpretation of human behaviour. Therefore, psychologists and educationalists are in need of the inferential statistics. It is hypothetical that all test scores and other measures of human behaviour are subjected to the occurrence of error. The errors may occur from a variety of sources. The way in which the sampling was planned and drawn, the limitations of the instruments, the procedures used in measurement, while drawing inferences about the behaviour of people in similar situations and also from different groups of people, non-representation of a specific group of the total population all may contribute to error variance. The errors occurring in these ways may affect a set of scores, the means, the standard deviations, the coefficient of correlations, the reliability and validity of tests and so on. But, however, statisticians provide methods of deriving the probability of errors to certain extent. Therefore, it is possible to estimate the level of errors occurring in any measurement. Statistical inferences are applied to all predictions in psychological and educational research. Statistical inferences help to predict the nature of results and also enable us to say how far the predictions are reliable. Different types of errors are discussed in statistical books. Every author discusses the same subject-matter in different way. They also use different subscript notations. This may cause confusion to beginner or to one who wishes to understand something about testing and measurements. There are also difference of opinions and disputes among the authors regarding some of the statistical values. One of the problems that is in dispute and also causing difference of opinions is the estimate of errors.

Types of Errors and Error Variance

The researcher faces problems regarding the estimation of errors in his studies. What type of error he has to estimate from his data? When we speak about the errors in psychological testing, right from the very beginning of the study of experimental psychology till the end we cannot avoid the mentioning of the errors. Generally the errors we come across are the standard error of the mean or sample, standard error of differences between the means, standard error of estimate, standard error of measurement, standard error of the difference for paired observations, standard error of coefficient of correlation, standard error of standard deviation, standard error of frequencies, proportions and percentages, standard error of median and centiles, and many others. Therefore, in treating the experimental data in a clear and meaningful way what type of inferential statistics and the error variance the researcher can apply, to interpret the results and write the discussion. The researcher after having planned his survey by selecting any one of the sampling methods according to his convenience and feasibility may collect data and treat the data with suitable descriptive and inferential statistical methods. The advantages and disadvantages of all the statistical methods are also stated then and there and therefore, he can estimate the errors of his study to establish the stability of his results. With this view the error estimates which are very much essential for psychological and educational researches are discussed here under various heads without going too much into the statistical equations and formulas. The estimate of errors discussed here may be used to test the accuracy of psychological and educational studies.

Sampling Errors

In the sampling methods described in the preceding pages, the important characteristics like the randomness, representativeness, the probability and the adequacy have been stressed if sample were free from bias or error. But always some kind of departures are taking place in sampling. As a result of this, errors in sampling are occurring and also the same become biased. Therefore, this type of errors called sampling errors. When the sample is not adequate and the sampling procedure is deviated then there is no confidence in the stability of the results and it also results in sampling error. Further, this is also leading to inconsistent results when different samples of persons are tested with the same instrument. The sampling error is different from the error of measurement which was discussed earlier. The sampling error is found to be often affecting the reliability of a measure. Gullikson (1950) while discussing the accuracy of test scores equates reliability with error of measurement and equates validity with error of estimate. The sampling is made to generalize the characteristic of the parent population. How far the result of a sample helps the investigator to generalize the whole population? The sample result is used to estimate the population characteristics.

Suppose, for example, the researcher wanted to know the level of mental abilities or the distribution of mental ability test scores of school students in a particular state in India. He cannot test all the students which may amount to more than 50 lakhs. Therefore, he limited the sample of the state to certain number, drawing the sample randomly from certain schools and considered it as the representative of the total student population of the state. But if he could test all the students studying in schools then there is no need for the sample or inferential statistics. And also if he could fully ascertain that the sample of the state is correct representative of the total student population in all respects then he need not go for the inferential statistics since he is fully convinced that the characteristics of the sample are identical with the characteristics of the whole student population in study. In such case there is no sampling error.

But, however, carefully, randomly and representatively he has drawn the samples from the population, the measured means, standard deviations and other measures of the samples are very rarely identical with the means and standard deviations of the parent population. This means that there is some error in the sampling. This may be due to the elimination of some elements of the population from the sample or some elements are included more over others. This is said to be biased and hence not random or representative and leads to errors. The investigator instead of including the numbered elements might have included his friends and relatives in the sample. Therefore, this sort of deviations from sampling techniques often result in investigations. These are the contributing factors to the sampling errors. Sometimes the investigator himself may not be knowing the correct variables interfered in his sampling and caused the errors. But the error exists whenever samples are selected instead of studying the entire population with the view of estimating the population characteristics with the sample measurements.

If the sampling procedures are not properly designed and followed in drawing the samples the sampling error often occurs. If the sampling is biased one, then all the characteristics of the population do not have equal and proportional chances of representation in the sample. Biased sampling means deviating from the random sampling procedure, selecting some favoured groups of people and including them in the sample and excluding the numbered elements. In such cases any conclusion drawn from the sample will not apply to the general population. This sort of biased sampling causes errors in the experimental situations.

Statisticians have devised methods of estimating such sampling or biased errors. One way of avoiding the errors is to use the *probability simple random sampling* and the *stratified or controlled random sampling* methods. But the other way of dealing with this is to estimate the error, as the error in sampling cannot be eliminated fully.

Standard Error of the Mean

Suppose, for example, a researcher is very much interested in testing the level of service motive of all the National Social Service (NSS) cadets in the state of Tamil Nadu by administering the service motivation test on all the cadets. But the number of cadets are 20,000 in all. Having realized the difficulties and the problems involved in testing all 20,000 cadets, hc decided to administer the test on a small sample of cadets. For which by the use of the table of random numbers he has drawn 5 per cent of the population according to simple random sampling method. The 5 per cent of the population makes 1000 elements. Therefore, he conducted the test on 1000 NSS cadets and estimated the mean as 60.0 and the standard deviation 6.0. The remaining 19,000 elements are not tested. At this stage can he estimate the population mean on the estimation of the sample mean? According to inferential statistics he can estimate it.

The sample mean and the standard deviation of the distribution are sufficient statistics to estimate the population mean. The mean of 60 one *SD* above the mean equals a score of 66 and the same mean one *SD* below the mean equals 54. According to the properties of normal probability curve when the total range of scores is divided into *SD* units from either side of the mean (like +1 *SD*, –1 *SD*, +2 *SD*, –2 *SD* etc.), certain fixed proportions of the distribution are comprised within each *SD* unit.

The normal curve area between the mean score and the score of one standard deviation above the mean occupy 34 per cent of the total area under the normal curve. The area under the normal distribution curve stands for the total frequency of the curve 34 per cent of the total scores fall between the mean and one *SD* above the mean and 34 per cent of the total scores fall between the mean and one *SD* below the mean. This means that 340 (34%) students have scores between 60 (the mean) and (plus one SD score) 66 and 340 (34%) students have the scores between 54 (minus one *SD*) and 60. This makes the normal curve symmetrical.

In the case of 2 *SD* units from the mean on either side a fixed proportions of area are occupied under the normal curve. Thus between the mean score 60 with two *SD* units equals 72 above the mean and the mean 60 with 2 *SD* units below the mean equals 48. This means that between the mean and 2 *SD* units 475 students (47.5%) have scores between 60 and 72 and 475 students (47.5%) have scores between 48 and 60. In the case of 3 *SD*'s from the mean 495 students (49.5%) of

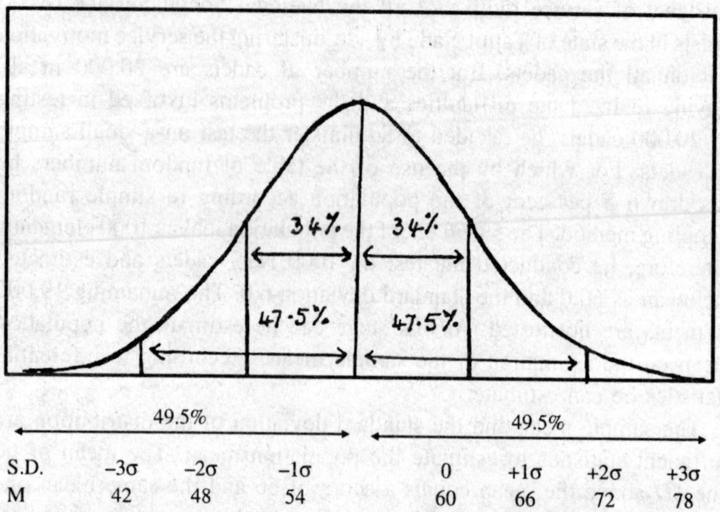

Fig. 15.1: Normal probability curve and estimation of population mean

the whole sample have scores between 60 and 78 and the same number of students (495) have scores between 42 and 60 (Fig. 15.1).

Therefore, the researcher can estimate the population mean in this way. Since the sample mean is 60 and the standard deviation of the mean is 6, it is possible to state that there are 68 per cent of chances for the true or population mean to fall between 54 and 66 and there are 95 per cent of chances for the population mean to fall between 48 and 72 and 99 per cent of chances for the population mean to fall between 42 and 78 per cent.

Thus, the researcher does not know the real population mean but he can estimate the probable range within which the true mean of the population falls in relation to the sample mean. How to find out this? One way of finding out this is to have different samples of the large population and to estimate the mean score of every sample and to find out the standard deviation of these means. Since large number of mean estimations are not possible a statistical method is evolved from the standard deviation of the single sample mean itself. It is the standard error of the mean. This is obtained by dividing the standard deviation by the square root of the number of cases of the sample minus one. The formula is as follows:

$$SE = \frac{SD}{\sqrt{N-1}}$$

Where SE = Standard Error of the mean

SD = Standard Deviation of the sample mean

N = Number of cases of the sample.

It is important to note that the number of cases of the sample is the determining factor of the size of the standard error. The larger the sample size the smaller will be the standard error. The smaller the sample size the greater will be the standard error and also the sample mean is more likely to deviate from the true mean and the dispersion score also may become wider. This infers the idea that if the representative sample is larger, then there is greater identity of characteristics with the population from which the sample is drawn. Therefore for a valid conclusion of any study the sample as far as possible should be larger and also should be representative and drawn randomly according to the sampling procedure.

This can be also observed that the standard deviation for a sample is smaller than the standard deviation of the population. Because the extreme deviations if any in population may be missing in the sample.

The standard error of the mean can be also estimated directly from the sum of squares of the deviations in the sample. The formula is:

$$SE = \sqrt{\frac{X^2}{N(N-1)}}$$

Where SE is the standard error of the mean

X^2 is the sum of squared deviations

N is the number of the sample size.

The necessity for estimating a standard error of the mean often arises in sampling studies with regard to the statistical stability. The standard error of the mean is the degree to which the obtained sample values fail to reflect the true population values. It may be also stated that a sampling error is the difference between the value of the mean calculated from a sample and that what have been obtained from the entire population. Therefore, the standard error can be defined *as the estimate of the sampling error*. The sampling error is generally stated as the standard error of the mean.

Errors of Measurement

In psychological measurements a basic problem that often destabilises the results is the problem of error. Errors occur in the process of measurement. Sheridan (1979) says that there is no way of avoiding some kind of errors while making measurements. We make errors in measurement and this is true in psychology and in all fields of social sciences. Though we cannot entirely eliminate errors we can reduce them to certain level and have our measurement as far as possible accurate. Therefore, it becomes more important for us to be cautious about the sources from which the errors are emerging.

One of the sources of error, is the measuring instrument. The measuring instrument itself may be having some defects within it. Every measuring instrument has certain limitation in discrimination. However, smaller the error from the correct measure the instrument makes, such smaller errors when accumulated become bigger source of errors. The precision of the measuring instrument needs to be tested.

Whether the errors are with the measuring instrument or with the persons who makes use of it there are two types errors often stated as errors of measurement. One is called systematic error and the other is called random error or unsystematic error. The random or unsystematic error is also called chance error.

Systematic Errors

However, carefully the measurement is planned there is some possibility of occurrence of these errors. These errors tend to occur in a consistent manner. If these errors are occurring consistently then the instrument (test) must be the main source and there may be errors in the correct fixation of graduations in the instrument or in testing the reliability and validity of the test. Because of this, the measurements made with this instrument cause the results systematically over-estimating of the true measurement or systematically under-estimating the true measurement. In the case of over-estimation, the errors may be consistently positive with positive signs and in the case of under-estimation the errors may be consistently negative with negative signs.

For example, we know the true weights of certain materials whose weights are already measured and indicated in them. These materials are now weighed on a scale one by one and the observed weights are recorded. Every material weighed now is having some difference with the true weight. Now, we can subtract the true weight of each material from the respective observed weight. The differences between these two values are called errors of measurement. This sort of errors may also occur in any measurement concerning human behaviour.

Unsystematic or Random Errors

The instrument may be accurately calibrated or the test might have been accurately validated with the view of reducing the systematic errors. But however, the instrument or the test is accurate the random or unsystematic errors will occur in measurement.

The positive errors and the negative errors stated above may occur sometimes frequently. The over-estimation and the under-estimation of the true measurement may be occurring with the same frequency. Therefore, both the errors may be of equally occurring. In such cases the positive errors are summed up separately and equated with the summed up negative errors. If these two types of errors are found to be approximately equal then we call them as random or unsystematic errors. Since both the positive and negative errors are equal when summed, the average of them becomes zero. The positive errors cancel out the negative errors. Random or unsystematic errors cannot be eliminated absolutely. However, carefully measurements are made, random errors may be accumulated in small measures and become one of the important errors. This is occurring by chance and therefore the random errors are also called chance errors. We can reduce these errors.

These types of errors are mostly estimated in psychophysical methods. For example, in the Muller-Lyer Illusion experiment the method of average error is used. In this experiment the participant is asked to equate the variable stimulus or judge the variable stimulus in comparison with the standard stimulus. In this process there is every possibility of judging the standard stimulus greater than the variable stimulus or smaller than the variable stimulus. While adjusting the variable stimulus either it may give a look that it is greater or smaller than the standard stimulus. Because of this there are both over-estimations and under-estimations causing in half of the trials positive errors and in the other half of the trials negative errors. The average of these errors results in zero, one cancels out the other. If there is any difference in this it may be called chance error. In this type of experiment the errors occur when the participant judges the standard stimulus using the variable stimulus as a measuring instrument. He sometimes over estimates the standard stimulus and sometimes under-estimates the standard stimulus using the variable stimulus as a measure. Hence, these errors also may be stated as errors in measurement.

Another psychophysical method, the method of minimal changes is also based upon estimating of standard stimulus by using the variable stimulus as measure. The over estimations and the under estimations are the estimations of errors only. In this method half of the estimations may be over estimation with plus signs and the other half may be under estimation with minus signs. The average of these errors will be zero. The positive errors cancel out the negative errors.

Circumstantial Errors

There is another classification of errors called circumstantial or ambient errors. These errors are occurring in the circumstances in which the measurement is made. These errors may also be called systematic or unsystematic or random errors. Unsystematic random error may be caused due to change in temperature, sounds and other situational changes. The formula for estimating the random error is:

$$e = X - X_t$$

Where e = a random or chance error

X = the observed score or measurement

X_t = the true score or measurement

The observed score or measurement is stated as: $X = X_t + e$. According to this formula we may expect that 'e' will be equal to zero. This is a theoretical notion, but in actual there may be a small difference between the two errors.

Hence, the difference between the true score (X_t) and the obtained score (X) is called an error of measurement. It is also assumed that the errors of measurement are uncorrelated with true scores. That is, the size of the error of measurement is not related to the size of the true score. If this occurs the variance of the obtained scores are equal to that of the variance of the true scores plus the variance of the errors of measurement. Therefore, the above mentioned formula can be also expressed as follows:

$$\sigma_o^2 = \sigma_t^2 + \sigma_e^2$$

Where, σ_o^2 is for the variance of obtained scores

σ_t^2 is for variance of the true scores

σ_e^2 is for the variance of the errors of measurement.

Standard Error of Measurements

The standard error of measurements has been a slow process of development in psychological testing. The importance of this value was realized only recently. But very rarely researchers make use of this measurement in their researches. Every psychological and educational measurement is subjected to some kind of error. However sophisticated instruments are used but errors in measurement are inevitable. Generally in testing, an error is stated as the difference between the true value and the observed value. This may be positive or negative. Guilford (1987) and Candland (1968) state that the sampling or standard error of the mean is different from the standard error of measurement. According to Candland "sampling error refers to inconsistent results when the same measure is applied to different samples of subjects", whereas the standard error of measurement refers to the inconsistent data when the same instrument is used on the same subjects for the second time. According to Ebel (1966) "the standard error of measurement is an estimate of the standard deviation of the errors of measurement associated with the test scores in set. The standard error of measurement is estimated by multiplying the standard deviation of the score by the square root of one minus the reliability coefficient." The formula is:

$$S_e = S_t\sqrt{1 - r_{tt}}$$

Where S_e = Standard error of Measurements

S_t = Standard deviation of the score

r_{tt} = Reliability coefficient

The standard error of measurement, according to Guilford (1978) is also called the standard error of an obtained score. The standard error of measurement, therefore, refers to the error variance in a measurement. For estimating the standard error of measurement we require two statistics, one the standard deviation of the distribution of obtained scores and two, the reliability coefficient of those scores. With these one can estimate the standard error of measurement using the formula given above. Let us work out one example for estimating the sample error or the standard error of the mean using the data given in Table 12.9. The formula is:

$$SE = \frac{SD}{\sqrt{N-1}}$$

In *Table 12.9*, the *SD* is 10.09 and the *N* is 52.

$$SE \text{ or } SEM = \frac{SD}{\sqrt{N-1}} = \frac{10.09}{\sqrt{51}} = \frac{10.09}{7.14} = \mathit{1.41}$$

This is stated as sampling error or standard error of the mean.

Let us work out another example for estimating the standard error of measurement using the data given in Table 6.1. The formula for estimating the standard error of measure is:

$$S_e = S_t \sqrt{1 - r_{tt}}$$

Here we are facing one serious problem in using the above formula to estimate the standard error of measurement. In any correlation we use two measurement of scores. The correlation is between variables *X* and *Y*. Therefore, there are two standard deviations, one for each measurement. But in the formula we are to use one standard deviation only. If we use the data given in Table 6.1 there are two standard deviations, 2.21 and 2.44 and the r_{tt} is 0.90. Guilford (1978) states that the $S_t\infty$ is a blanket index and it implies approximately equal discriminating power throughout the scale. The assumption is that both the *SDs* are equal approximately. Therefore, we can use any one of the standard deviations. But if there is difference between them it is better to use the Standard deviation which can give better accuracy. According to our data the standard deviation 2.21 may give us better accuracy. Therefore, the S.D. 2.21 is applied for estimating the standard error of measurement.

$$S_e = 2.21\sqrt{1 - 0.90} = 2.21\sqrt{0.10} = 2.21 \times 0.316$$

$$= 0.698 \text{ or } \mathit{0.70}$$

The standard error of measurement is 0.70.

This means that two-third of the scores or 68 per cent of cases fall in between the *SEM* of – 0.7 and + 0.7 and 95 per cent of cases fall between – 1.4 and + 1.4 *SEM*. That 5 cases out of 100 will have the *SEM* of – 1.40 or + 1.40.

Thus, we have shown how the standard error of the mean or the sampling error is different from standard error of measurement. These two are different measures. The errors of measurement include all the facts that affect the scores. The standard error of measurement gives an indication of the absolute accuracy of the test scores.

Another important dispersion of measurement is the *standard error of standard deviation.* Just like the mean, the standard deviation also varies or fluctuates from sample to sample. In a smaller sample, less than 100 the distribution is skewed to make it close to the normal. But with bigger sample we can make inference about the standard deviation. If the *SDs* vary from sample to sample then we can estimate the *SE* of the *SDs* by using the formula.

$$SES = \frac{S}{\sqrt{2N}}$$

Where SES = Standard Error of standard deviations

S = Standard deviation

N = Number of cases

Applying this formula to any data with several standard deviations we can estimate the standard error of the standard deviations.

PART III

STATISTICS IN PSYCHOLOGICAL MEASUREMENTS

INTRODUCTION

Though these two statistics (descriptive and inferential) may fulfil the requirements of the researchers, there are certain statistics which are specially important for behavioural and social sciences as the studies are based upon certain valid instruments. Psychological studies or measurements are generally made by using the validated test materials, and precise instruments. Psychologists always make it a point to test the instruments prior to the final study or investigation, for which they use the statistics of reliability and validity of measurements. They hesitate to make use of a test in their studies, unless they are fully satisfied with the test. The test should be reliable and valid.

There are several methods to test the reliability and the validity of the tests. The most widely used methods of reliability are the test-retest, the parallel forms, and the split-half. In validating a test they use the correlation methods between the scores of two similar tests or two parallel forms and also apply the Spearman-Brown Formula for enhancing the level of reliability. Similarly there are several methods for validating the tests.

In certain research works the descriptive and the inferential statistics could not be used in their original forms. In such cases the researcher converts the test scores into norms on the basis of important independent variables and also converts the scores into percentile ranks and scores. Psychologists are not satisfied with these statistics and they went on further to explore the factors that cause change in the results and they found that the factors that influence change in behaviour fall under three categories, such as, the general factors, the specific factors and the group factors. Therefore, a statistics called factor analysis has emerged for identifying the factors that influence the change in the results. This is highly sophisticated psychological statistics, generally not found in the statistical text books.

16

RELIABILITY OF MEASUREMENT

Introduction

If any one wants to measure some aspect or quality of a group of people or individual then he looks out for a best test available for his purpose. A researcher who works seriously for his Ph.D. Degree in the area of attitude measurement, a teacher who wants to measure the mental abilities of his class students, an industrial Psychologist who is engaged in selection and placement of the employees, a clinical psychologist intended to assess the causes of the maladjustment of a child, the defense psychologist who wishes to reveal certain special abilities in the Air Force men—all seek to find out suitable test that could serve the purpose. There are quite a number of tests available in every area and which test one has to choose. Also there are different types of tests such as long tests and short tests, new tests and old tests, familiar test and unfamiliar tests. Each of these test producers claim that his test is genuine. Therefore, he is in a confused state in choosing a correct test. But every test has certain virtues and no one test has all the virtues. A test maker could not put all the desirable factors in his test. Further, there are also several testing procedures. Therefore a test must be selected for a particular purpose for which it has to be used. The test user also should evaluate the new developments in developing tests and the uses of the tests. Therefore, the test user must choose among the tests the right tests and know the testing procedures, not only the best procedure but also how well it satisfies his needs scientifically.

Every scientific research in the field of psychology and education is an attempt to measure some characteristics of an individual or a social group. In this attempt a psychologist really confronts with a problem of selecting a best instrument for his purpose and the procedure which would give him a successful result, also the ground on which he can get the correct result. For which he must evaluate and justify the test. There are several test evaluations for this purpose. Therefore, it is better to discuss the two important aspects of the tests, the reliability and validity of measures.

Importance of Reliability

The concept of reliability is a crucial problem for the test constructors to describe the outcome of the measuring process and measuring instruments. What meaning the measurement experts and test designers ascribed to it seems to be similar to the meaning ascribed by the common sense world. The meaning of reliability as conceived by the measurement experts cannot be the same or similar to the meaning prevailing in the common sense world. For example, I say that my watch is reliable. It means that it gives correct time, because it consistently agrees with the wall clock in showing the time. To this extent my watch is reliable. The wall clock and my watch may be, say, 15 minutes fast or slow to the standard time. But this does not affect the reliability. All that matters is the two separate measures (watch and clock) of the same thing (showing time) yield closely similar assessment, in terms of reliability. So also a watchman in an office is said to be reliable. What does it mean? The watchman goes to the office in time, stands at the gate for the whole day and leaves the office at the time when the office is closed. Suppose the watchman goes to office every day one hour late, sits there and sleeps whole day and leaves the office every day one hour earlier. He is reliable because he is consistent in his behaviour from time to time. According to the measurement expert's approach both my watch and the watchman are reliable. The behaviour is reliable, but not desirable in both cases. Similarly, a test may yield reliable information, but not suitable or desirable for the particular purpose for which we need it.

What is Reliability?

Reliability is defined in various ways. Every author gives his own definition. But generally reliability stresses the degree of agreement between two equivalent measures of the same variable. Guilford (1954) provides a logical definition of reliability. According to him the reliability of a set of measurements is defined as the proportion of true variance. Any test score obtained from a population may have two components such as true variance and error variance. Reliability refers to the proportion of true variance. In other words absence of error variance. Reliability is always expressed in terms of 'reliability coefficient' (r_{tt}). It is stated as the self correlation of the obtained test scores.

Reliability is a statistical problem often arises in psychological measurements. It refers to the accuracy of a measurement and tries to see to what extent the test is reliable. Reliability may be also defined as the degree of consistency with which a test measures whatever it measures. In other words, it is the unit of absence of errors. It cannot be directly measured. It is mainly concerned with the amount of consistency in all its measurements. If the measuring stick is correct we must try to find out the error in the sampling. But however, we must see the error both in the measuring stick and in sampling or in the method of measurement. It is this that we are discussing under the title of reliability. Reliability in a measurement is higher when the error in the measurement is very low. The validity of a test depends upon the reliability of its measurement. But reliability does not always determine the validity.

Sometimes there may be certain amount of lacking of reliability in testing. We may find certain uniformity in our findings when we measure a mental ability of a group in a test. In administering the same test to the same subjects under the same condition we may find the same result rarely. We may not be also expecting the same. But if the two sets of measurements closely correspond to each other we are satisfied.

The reliability of a test is determined by the correlation between the two sets of scores obtained from the same subjects at two different times. If the two sets of scores correlate perfectly the reliability is perfect. If there is no correlation the reliability is zero. The coefficient of correlation obtained in this way is known as the coefficient of reliability. If the coefficient of correlation between two sets of scores falls below 0.6 then the test is not valued scientifically. Approximately the reliability should reach 0.8 or 0.9. Therefore, a measurement can be said to be reliable if an individual remains nearly the same in the retest measurement and that the standard error of measurement is low and the coefficient of reliability remains higher. Now we may ask how accurately the test measures whatever it measures. What is the precision of the score and how accurately it will be reproduced if we measure the sample again? Suppose we are testing a group of boys in the school today with an aptitude test. Again on the next week or after three weeks we test the boys with the same test. When we compare the two scores for the boys we may find that they are seldom exactly the same. The two test scores differ. But most of the differences among the boys will be considerably small and some may be somewhat large and in few cases the differences may be much greater. This means that one measure is not perfectly reliable as a measure of the boys' aptitudes. The results seem to be inconsistent to some extent from first week to third week. Why this kind of difference? Several reasons have been attributed to this type of differences.

Thorndike and Hegan (1970) state three kinds of reasons for this type of inconsistency between the first test and the second test. According to them, "the person may actually have changed, the task may have been different, and the limited sample of behaviour." But these reasons are not applicable to all types of tests as they state. However, time is a determining factor of human behaviour. In our example on the first day when we tested the persons they might have been more careful and restful than on the day when the second test was conducted. They might have learned many things during the course of 2 or 3 weeks time. Some physical changes also might have taken place. Also there might have been some changes or development in their mental abilities, interests, learning, health, moods and temperament. Therefore, the boys' performance on the same test during the course of 2 or 3 weeks time also may change. It is possible that scores may vary from one trial to another and no one measure can be trusted absolutely. The scores obtained in a test indicate roughly the level of ability or typical behaviour of a person. The small the standard error of the scores the more precise the judgment is. Reliability information is the indication of the level of confidence we can have on a measurement. Reliability always refers to the consistency of a series of measurements. This can be observed in various ways. For example, the same test can be repeatedly administered or a series of parallel forms may be used. Further, these tests may be administered with different experimental procedures and measures have different types of variations and, therefore, there may be different reliability coefficients.

Reliability is generally expressed in terms of reliability coefficient. This is the correlation between the scores of two measurements obtained in the same manner. In any measurement there is individual variation and individual variation is described by standard deviation. The variation represents the combination of differences of the individuals. In any two measurements the obtained scores may vary to some extent, but the true scores may remain constant from measure to measure, because a person's true ability is constant. We may determine the person's score on a very large sample of behaviour. The true score is based on this. The true score becomes the average score the person obtains on a very large sample or series of samples. The error score is the variation or fluctuation of scores in the person's performance in the large samples or series of samples. The size of the sampling error is stated by the standard error of the measurement. Therefore, the obtained score is the combination of the true score and the error score in a measurement. These variances have direct influence in the correlation between the scores obtained from two measurements.

While discussing the reliability measures it is also important to look into the length of the instruments. A single question or a few number of questions will never reveal a person's ability. By adding more and more questions on the sample problem we may be making a good estimate of one's ability or intelligence. The longer tests not only may reveal a person's ability to a greater extent, but also reduce the influence of other factors. When we increase the number of items in the test for lengthening it is important that the nature of the problem of the test should not be changed. But at the same time we must be cautious in lengthening the test. Too many questions or items in a test may make the participants to lose interest in answering the questions. And also too many alternatives to answer a question or item may lead the participants to spend more time to guess and this may reduce the accuracy. All these may contribute to reduce the reliability of the test. A reasonable increase in test length may have greater effect on its reliability, but it has smaller effect upon validity of the test.

Generally, we assume that reliability determines validity. How far reliability determines or limits validity? The rule is that correlation between the test scores and an independent criterion scores should not go higher than the square root of the correlation (reliability) between the two forms of the test. Suppose, for example if the reliability of a test is 0.75 the validity should not exceed 0.86. Black and Champion (1976) state that reliability is a prerequisite for the validity of a test. A measuring instrument should be demonstrably reliable for determining the validity. The measuring instrument must reflect some attribute consistently. Otherwise, it will not be considered as a valid instrument of that attribute. Suppose, for example a person is believed to be authoritarian in his personality make up. To ascertain this the 'F-scale' was administered on him several times and there were widely different scores in the measurements. Therefore, the measure of authoritarianism was regarded as unreliable. It can be also stated that an individual is not influenced by environmental factors or extraneous variables from one test administration to another, but significant score differences were observed in the measurements. The more score differences in the measurements the more unreliable the test is. The first measurement and the subsequent measurements of the same test on the same sample should not show significant differences among them.

Any measuring instrument can never be subjected to the estimation of reliability by using the responses of a single individual to that instrument. Responses are mostly collected from a large aggregate of individuals of the same type for estimating the reliability of a measuring instrument which was specially designed for this population. Suppose, for example if a researcher wanted to study the reactionary—radical tendencies of engineers employed in government departments then he will have to pretest his scale or questionnaire on a separate sample of engineers and validate it. This pretesting has to be done only among the engineers and not on any other occupational group. Information from other occupational groups for the improvement of the instrument designed for the engineers will be of very little value. But critical views and suggestions from other engineers for the improvement of the instrument, the researcher should take into consideration.

Methods of Determining the Reliability of Measures

There are several methods for estimating the reliability of the measuring instrument. These methods may be classified into two categories. They are: (1) external consistency procedures and (2) internal consistency procedures.

According to external consistency procedures the test scores of one measure is compared against the same test scores obtained on another occasion. That is, the same instrument administered on the same population on two different occasions and the scores are correlated to determine the reliability of measures. Or another form of the same test designed is administered on the same population and the two test scores are correlated for estimating the reliability. There are two important methods based on external consistency procedure. They are: (1) Test retest method, and (2) Parallel forms of the same test.

Test-Retest Method

Among the two external consistency procedures the test-retest method is more often used by the researchers for estimating the reliability of a measuring instrument. An attitude scale that is designed according to the procedure to measure some aspect of social issues is administered on a group of university students located in some area at a given time (t_1). After a period of four weeks the attitude scale is again given to the same students of the same university (t_2). These two sets of test scores of the attitude scale are correlated and resulting correlation coefficient is the reliability measure of the attitude scale. This may be stated as the reliability coefficient. The test-retest method of reliability coefficient is estimated by administering the same test on the same population on two different time periods. In this method we expect that a reliable measure is to reflect the characteristic to the same degree and no extraneous or environmental variables interfere with the test scores and increase the score differences significantly. A correlation coefficient obtained according to this condition between two test scores for the same population is accepted as valid. The researcher, sometimes, may wish to use the instrument in an experimental situation to see whether some experimental variables can interfere and modify certain attitudes and affect the reliability of the instrument. In such case he can use his instrument that has been already pretested and accepted as valid. If the researcher finds some change in the scores on some measured characteristic (attitude) in an experimental situation, he can assume that some experimental variable has interfered in the measurement and caused the change and the change was not due to any random variable operating in the situation. The test-retest reliability method is found to be more useful in stable situations where there is no environmental change significantly from one test administration to another.

The test-retest reliability method has certain advantages. This method allows the test user to compare itself directly. The same test score is used to verify the same test. Sometimes, the instrument performs unreliably in a test-retest situation. In such case the instrument may require item by item analysis to find out which are the items that discriminate between individuals having different amount of the trait that is studied and which are the items that do not discriminate between individuals. The reliability of the instrument may be increased by eliminating the items that do not discriminate. This kind of item analysis is also a kind of testing the internal consistency of the test. We are discussing this at the end.

In test-retest reliability method we conduct two measurements with the same instrument on two occasions. Irrespective of the time period it makes the measurement as continuous one. In other words there is continuity in the measurement. This method is very easy to test the reliability of an instrument. We use one instrument on the same population and we get two scores for our purpose within a very short period. Thereby, it is quick and saves time. The estimation of reliability is also easy. Another advantage of this method is that the extraneous or environmental factors are controlled in this method to the maximum possible extent so that the measurements are not diluted by error variances.

The test-retest reliability method also has certain disadvantages. The criticism is based upon the administration of the same test on the same population two times with short period of interval. The individuals who have answered the test items few days ago are able to recall how they answered to the items of the instrument first time. They will show the same reaction to the items and repeat the same answer to be consistent with their first response. This kind of bias is quite common. If the individuals recall their previous responses and answer the items with the same manner then the reliability of the instrument is not correctly attributed to the instrument. But we cannot destroy their memory or change the items. The only possible way of avoiding this sort of situation is to lengthen the time period between the first administration and the second. There is no definite time period mentioned by the test experts for this. However, a period of 3 to 4 weeks interval is regarded as reasonable time.

The test-retest reliability method is not free from defects. Having too much faith in this method, test users apply the instrument to a large variety of target groups. It is better that they must understand the limitations of their method. The reliability should be ascertained on every sample before the instrument is used for collecting data for a research purpose.

It will not be possible for the test user to understand the impact of the extraneous or environmental factors on the individuals who are taking part in the test-retest reliability test. A significant correlation coefficient will not give sufficient guarantee that the reliability of the instrument is highly valid. It may be due to chance success. Therefore, the test user should constantly subject his test to experimental verification for each population he studies. Further the reaction of the individuals who take up the test for the second time may vary when they see the researcher entering the testing situation for a second time. This may have some effect upon the responses.

Parallel Forms of the Same Test

Another important external reliability method is the parallel forms of the same test. When an investigator wants to estimate the reliability of an instrument by the method of parallel forms, he designs two tests of the same type, but with different items to measure the same trait or phenomenon. Several test experts in the past have followed this procedure. For example, Cattell (1950) devised a test called 15 Personality Factor Inventory with Form A and Form B as parallel forms. Both Form A and Form B measure the same personality traits equally. It is also possible in this case to use test-retest method for estimating reliability of the instrument and also the influence of experimental variable can be determined. Form A of the Inventory may be administered first and Form B secondly after few weeks. This will solve the effect of memory and recall problems and the influence of bias responses in the test-retest method. Since, there were different items in both forms there is no chance for the respondent to recall the items seen in the first form.

The parallel forms method is also called alternate forms method. The two forms for this method should be developed in content, style and statistical characteristics. Each form should have the same number of items. The means that the standard deviations of the scores of the two forms also should be about the same. Another procedure of having parallel forms is that we can construct a test with 120 items of the same type for assessing a particular problem and divide it into two forms, by assigning 60 even numbered items to one form and 60 odd numbered items to the other form. Any one method can be adopted to have parallel forms. But this sort of division will not give guarantee that the two forms are equal or parallel in all characteristics. If the two forms of the tests really equal or parallel then the reliability may be estimated by administering one form first and then the other. There is a way of checking this. Suppose, for example, we are assessing the reliability of the test by administering it to a group of 400 university students. The first form may be given to the first 200 students and the second form may be given to the other 200 students. The scores of the 200 students from the first form are correlated with the scores of the other 200 students from the second form. One should bear in mind that a correlation of +1.00 is positive correlation which represents perfect agreement between the two forms, –1.00 is the negative correlation which represents perfect disagreement between the two forms and a zero correlation means no correlation and no agreement. In assessing reliability of measuring instrument the correlation coefficient should not be less than +0.60.

Thorndike and Hegan (1970) state that the use of two parallel test forms provides a very sound basis for estimating the precision of a psychological or educational test. One disadvantage with parallel forms is that the parallel should be available. Generally a second test of the same is very rare. Secondly, the time to administer the other form on the same sample. So much time may not be available for a second testing. This difficulty can be overcome by following one of the methods suggested earlier.

Internal Consistency Measures of Reliability

In the measuring instrument the items that assess the same phenomenon should have some logical consistency within it. Internal consistency refers to agreement within. A test possess high internal consistency if it is composed of items which all measure the same thing and which are, therefore, highly inter-related. A measure of internal consistency provides one measure of test reliability. In test development procedure it is important to check the internal consistency of the test. Items which are not consistent or in disagreement with other items are removed and thereby the internal consistency is improved. A test with internal consistency increases the chance of the test being reliable. Internal consistency of a measuring instrument is determined by using (1) the split-half technique and (2) an item discriminative analysis.

The Split-half Technique

This method is very often used for estimating the reliability of a test when one single test is applied to a single group. Sometimes it may not be possible to administer one test on two different occasions on the same group or individuals. In such case a test consisted of 60 items may be administered on a sample. Then the test may be divided into two halves each having 30 items, thereby each one considered as a separate test. The division is made on the basis of odd-even items. The odd numbered items constitute one test and the even numbered items the other test. Thus, the full test is divided into two (30 + 30) equal independent test. Suppose, for example, if the full test consisted of 61 items instead of 60 items, then the problem for the test user is where to put the 61st item since it is an extra item in the test. The principle is to put the score of the 61st item on both half tests. Thereby each half test will have 31 items. This is called split-half method. The odd numbered item test and the even numbered item test are scored separately. A correlation is computed between the scores of these two independent tests. The higher the correlation coefficient the more the internal consistency the instrument has. According to the rule a correlation of 0.90 or higher is regarded as the indication of higher internal consistency and reliability of the instrument.

The Spearman-Brown Formula

According to this method one test is split into two halves and splitting a test like this means that the scores, whose reliability is determined have been obtained from half-length tests and hence the value cannot be directly applicable to the full length test. In general the larger the sample of a person's behaviour we have, the more reliable the instrument will be. If the two halves of the test are equivalent by actual correlation then we can get an unbiased estimate of the total test reliability from the correlation between two half tests. Therefore, to obtain the internal consistency reliability of the total test, it becomes necessary to step up the half test correlation to the expected full length value by applying the Spearman-Brown Prophecy Formula.

The Spearman-Brown formula is: $r_{tt} = \frac{nr}{1 + (n-1)r} = \frac{2r}{1+r}$

Suppose, for example, the internal consistency of a test is estimated as 0.80, the reliability coefficient of the test is assessed by applying the Spearman-Brown formula here. Accordingly, the reliability coefficient of the test is:

$$r_{tt} = \frac{2r}{1+r} = \frac{2(0.8)}{1+0.8} = \frac{1.6}{1.8} = 0.88$$

r_{tt} = reliability coefficient of the whole test.

r = correlation between the two halves

It is obvious that the Spearman-Brown formula enhances the internal consistency reliability of the test from 0.80 to 0.88.

The Kuder-Richardson Formula 20

A number of techniques have been developed for assessing reliability of a test with one administration and using the data for reliability estimation. Kuder and Richardson (1937) have published an article, "The Theory of Estimation of Test Reliability" which included a series of formulas. Two of them were widely accepted as a basis of estimating test reliability. They are Kuder-Richardson Formula-20 and Kuder-Richardson Formula 21. These formulas have Spearman-Brown Formula as the basic format. Let us state them here for those who are interested in such statistics.

The Kuder-Richardson formulas provide reliability coefficients of a single test from a single administration. Generally the required information is the number of items in the test, the standard deviation of the test scores, the difficulty of each item in the test or the average difficulty of all items as reflected in the mean test score. It is statistically sound and convenient for test users and therefore the Kuder-Richardson formulas are now widely used for estimating the reliability.

The formula 20 is: $\nu = \frac{K}{K-1}\left(1 - \frac{pq}{\sigma^2}\right)$

In this formula K represents the number of items in the test. p stands for proportion of right responses, q stands for the proportion of wrong responses (responses not correct). In this equation p plus q always equals 1.

Σ means the sum of the proportion of right responses (p) and the proportion of wrong responses (q)

σ^2 = Standard deviation of the scores of the test.

According to this formula the proportion of correct responses to each item be determined and in each case multiplied by the proportion of responses which are not correct. These p time q values for each item are then added for all items. Then the sum divided by the square of standard deviation of the test scores and subtracted from 1 and then, is multiplied by the fraction K over K minus 1 to obtain an estimate of the reliability of the test scores. Usually the Kuder-Richardson formula-20 is more suitable for tests in which the items are scored with 'yes' or 'no' answers and giving one mark to correct answers and no mark to wrong answers. Kuder-Richardson formula-20 provides a coefficient for any test which is equal to average of all possible split-half coefficients. The Kuder-Richardson coefficient is considered as good estimate to an equivalent form of correlation.

The Kuder-Richardson Formula-21

The other Kuder-Richardson Formula-21 is somewhat less accurate. But it is very simple to compute *KR-21* coefficient. The information required for this formula is the mean (M), the number of items (K), and the standard deviation (σ^2). This formula provides nearly the same result for most tests as the formula *KR-20*. But the coefficient is little lower. This formula is stated in different subscript notations.

One way of stating Kuder-Richardson formula 21 is:

$$r = \frac{K}{K-1}\left(1 - \frac{M\left(1 - \frac{M}{K}\right)}{\sigma^2}\right)$$

The other way is:

$$r = \frac{K}{K-1}\left(1 - \frac{M(K-M)}{KS^2}\right)$$

S = Standard deviation of the test scores.

This kind of internal consistency procedures are not applied to speed tests, because the parts of the tests in speed tests are not independent. The correlation of test scores obtained in one administration is always higher than the correlation of test scores obtained in two separate administration. Therefore, the split-half reliability coefficient or the Kuder-Richardson reliability coefficient is always high but not very much accurate.

I. Every author or test user who uses Kuder-Richardson formula-21 states it in his own way. But the original form as given by the authors is:

$$KR_{21} = \left(\frac{K}{K-1}\right)\left(1 - \frac{\overline{X}(K-\overline{X})}{KS^2}\right)$$

Where K = stands for the items of the test, scored as either right (1) or wrong (0)

$\overline{X}$ = the mean score of the test

S^2 = the variance of the test scores (SD).

II. But the formula as stated by Cronback (1966) is:

$$\gamma\, KR\text{–}21 = \frac{K}{K-1}\left(1 - \frac{M(K-M)}{KS^2}\right)$$

Where M = mean

K = the number of items in the test

S = the standard deviation

III. The formula *KR* 21 as stated by Ebel (1965) is :

$$r = \frac{K}{K-1}\left(1 - \frac{M\left(1 - \frac{M}{K}\right)}{\sigma^2}\right)$$

Where K = stands for number of items in the test

M = mean

σ^2 = Square of standard deviation

IV. The same *KR*-21 formula is stated by Thorndike and Hegan (1970) as given below:

$$r_{tt} = \frac{n}{n-1}\left(1 - \frac{M_t\left(1 - \frac{M_t}{n}\right)}{S_t^2}\right)$$

Where r_{tt} = is the estimate of reliability

n = number of items in the test

M_t = the mean score of the test

S_t = the standard deviation of the test

V. The equation of Kuder-Richardson formula-21 as stated by Guilford (1987) is:

$$r_{tt} = \left(\frac{n}{n-1}\right)\left(\frac{S_t^2 - n\,\overline{p}\,\overline{q}}{S_t^2}\right)$$

Where r_{tt} = reliability coefficient
n = number of items in the test
$\bar{p}$ = average proportion of passing examination in all items
$\bar{q}$ = average proportion of failing examinees in all items
S_t^2 = square of standard deviation

The reliability coefficient estimated according to $K - R$ formula-21 is lower than the $K - R$ formula-20, but it is higher than the estimate of odd-even method. The $K - R$ formula-21 is having one advantage over the formula-20. It needs only few data. We need not make item score counts. Guilford has simplified the $K - R$ formula-21 and has given a formula alternate to the formula. It is:

$$r_{tt} = \frac{nS_t^2 - \bar{R}\,\bar{W}}{(n-1)\,S_t^2}$$

Where r_{tt} = reliability coefficient
n = number of items in the test
$\bar{R}$ = average number of right responses. It is the mean of the total scores. Total score is the number of right responses.
$\bar{W}$ = average number of wrong responses $(n - \bar{R})$
S_t^2 = square of standard deviation

In all the equations shown above certain subscript notations are common to all. But only in some equations the subscript notations vary.

It is important to note that all Kuder-Richardson formulas are used to estimate the internal consistency reliability of a test administered only once on a sample. The internal consistency reliability is mostly applied to homogeneous tests i.e. having the same characteristics in every respect. The items of the test measure the same trait to about the same level. When the item inter-correlations are high the internal consistency reliability is also high.

Item Discrimination Analysis

Another method of approach for estimating the reliability of a test internally is through item discrimination analysis. When we design a test with about 30 items we cannot expect that all the items are having the same characteristics to make the test a homogeneous and internally consistent. In its analysis the test designer spots out inconsistent items and removes them. Secondly he traces out the items that fails to discriminate between individuals who score consistently higher from those who score consistently low and either improves the items or removes them.

Suppose, for example, a researcher has administered a family planning and birth control attitude scale on 200 people. The attitude scale contains 30 items of which 15 were positive items and 15 negative items. It is a cumulated rating scale of Likert Type Pattern. It is a five-point scale and each item is answered with five categories, like strongly agree, agree, undecided, disagree and strongly disagree. Weights were assigned to the responses from 5 to 1 (5, 4, 3, 2, 1) for positive items and 1 to 5 (1, 2, 3, 4, 5) for negative items. In this method a person who is positively inclined towards the problem may choose strongly agree response for which the weight is 5. The same person is expected to respond to negative item with strongly disagree response for which the weight is 5. This person, therefore, is expected to score on this scale $30 \times 5 = 150$. The maximum score in this scale is 150 and the minimum score is $30 \times 1 = 30$. Hence, the scores range from 30 to 150.

Generally, a person with very high total score would be inclined to choose the responses which are assigned with high scores like 5 or 4 in positive statements or items (strongly agree or agree) and 4 or 5 in negative statements (strongly disagree or disagree). A person with a very small total score would be inclined to choose either disagree or strongly disagree responses of the positive items for which the weights assigned are 2 and 1 and also would choose strongly agree or agree responses for the negative items for which the weights assigned are 1 and 2. In an investigation it is also possible that some respondents who give consistently responses weighted with a 4 or 5, respond to certain items weighted with either a 1, 2 or 3. Sometimes, it is also true on the other way that individuals who have consistently provided responses weighted with a 1, 2 or 3 respond to certain items weighted with either 5, 4 or 3. This means that some individuals deviate from the correct way of responding to items, which is stated as inconsistencies. If there are many items which were answered inconsistently they are interfering with the internal consistency of the test and the reliability coefficient would be unreliable. These items are not from the same universe as the other items. These items, therefore, should be removed to improve the internal consistency and the reliability of the test.

An Illustrative Experiment to Identify the Inconsistent Items

The family planning and birth control attitude scale consisted of 20 items, 10 positive items and 10 negative items was administered on 40 university students. It is a five-point scale and the respondents were asked to answer with five responses such as strongly agree, agree, undecided, disagree and strongly disagree. The responses were weighted with 5, 4, 3, 2, 1 for positive items and with 1, 2, 3, 4, 5 for negative items. Thereby, the highest possible score is 100 and the lowest possible score 20. The scores ranged from 20 to 100, the low scores indicated unfavourable towards the family planning and birth control. The questionnaires were collected from all 40 students and scored according to the procedure and the scores for all 20 items totalled for each individual. Then the scores of the 40 students were arranged in rank order, from highest to lowest. Then those who fall above Q_1 and those who fall below Q_3 were selected. That is those who fall above upper quartile and below the lower quartile were compared. This is to find out items which will consistently separate those people who were "high" from those who were "low" in the attitude continuum. In this procedure each item is separated to a measurement of item ability to separate the highs and lows. This is called discriminative power of the item. Thus, we have 10 respondents in the upper group and 10 in the lower group and 20 in the middle group.

The individuals whose scores fall in the centre are 50 per cent in the distribution. These scores are excluded from further consideration according to the procedure of the internal consistency. The logic is that the discriminative power of the item is clearly observable in the case of extreme attitude scores. Now we have the scores of two extreme groups, the upper and the lower. The experimental data in Table 16.1, clearly illustrate the principle and the procedure of estimating the discriminative power of each statement or item. We understand from the data as given in Table 16.1 that the lowest discriminative power of the statement is 1.4 for item No. 9 and the highest is 4.0 for the item No. 1. The mean scores (the discriminative power) of the statements or items discriminate between those who possess the trait to a high level and those who possess it to the low level. It is better to choose the items that have higher mean or discriminative power to constitute a test. At the same time the items that have low mean or discriminative power may be improved by observing the scores in Table 16.1. For example item

No. 9 is responded with weights 4 and 3 in the upper group and 3 and 2 in the lower group. There is not even one score with 5. All weights are 4 and 3. Actually the score 3 should not be there. There are 3 such scores in the upper group and 3 scores in the lower group indicating that three respondents in the lower group gave undecided answers to the item. This means that this item is not providing clear cut ideas to the respondents to respond with definite acceptance or rejection of the trait. Therefore, this item needs improvement or it should be removed. Similarly items 12 and 14 are also responded with undecided answers by the respondents. Two respondents for item 12 and one for item 14 have given deviated responses (3) in the upper group.

They are expected to answer with response weights 4 and 5. Similarly in the lower group there is one response in item 12 and one in item 14 with response weight of 3, instead of 2 and 1. This means that there is some ambiguities in these items which misguide them to respond with the answer undecided. The mean score or discriminative power of these two items is the same as 2.1. There are no deviated responses in other items and, therefore, they are construed as reliable items.

There are certain reasons why some items fail to discriminate between the highs and lows. These items may be ambiguious and failed to yield a clear cut idea to the respondents. Therefore, these items do not measure the trait under study. The researcher might have assigned the response weight inappropriately to these items. By making suitable corrections in the items they may be reinducted in the scale. The important function of item discrimination analysis is to make improvement in the reliability of a test by eliminating those items that are inconsistent with the rest of the items and fail to discriminate between the individuals possessing the trait in varying degrees. It is always safe and better that the researcher should include large number of items in his test, more than 20 items if he were to use item discrimination analysis. On completing the item analysis several items or statements will have to be eliminated from the test because of their failure to discriminate between individuals who possess the trait in varying degrees.

The researcher also should see that the range of scores should not be narrowed down too much by eliminating too many statements or items in the list. This may affect and restrict the statistical treatment of the data. But at the same time if the researcher retains too many items there will be increasing chances of including items that have

Table 16.1 : Showing the item discrimination analysis and the discriminative power of the items

Group	*Total score for 20 items*	*Scores for each item* 1	2	3	4	5	6	7	8	9	10	11	12	13	14	15	16	17	18	19	20
Upper 1	95	5	5	5	5	4	5	5	4	4	5	5	5	5	4	4	5	5	5	5	5
2	95	5	5	5	5	4	5	5	5	4	5	5	5	5	5	4	4	5	4	5	5
3	94	5	5	5	5	4	5	5	4	4	5	5	5	4	5	4	5	5	4	5	5
4	94	5	5	5	5	4	5	5	4	4	5	5	5	4	5	4	5	5	4	5	5
5	93	5	5	5	5	4	5	5	5	4	5	5	4	5	4	4	4	5	4	5	5
6	93	5	5	5	4	4	5	5	5	4	5	5	4	5	4	5	4	5	4	5	5
7	92	5	5	5	5	4	5	5	4	4	5	5	4	5	4	5	4	4	4	5	5
8	91	5	5	5	4	4	5	5	4	3	5	5	4	5	4	4	4	5	5	5	5
9	90	5	5	5	4	4	5	5	4	3	5	5	3	5	4	4	4	5	5	5	5
10	88	5	5	5	4	4	5	5	4	3	5	5	3	5	3	4	4	5	5	4	5
Total for upper		50	50	50	46	40	50	50	43	37	50	50	42	48	42	42	43	49	44	49	50
Lower 1	38	1	1	2	2	2	2	2	2	3	1	2	2	1	3	2	2	2	2	2	2
2	38	1	1	2	2	2	2	2	2	3	2	2	3	1	2	2	2	2	2	2	1
3	37	1	1	2	2	2	2	2	2	3	1	2	2	1	2	2	2	2	2	2	2
4	36	1	1	2	2	2	2	2	2	2	1	2	2	1	2	2	2	2	2	2	2
5	35	1	1	1	2	2	2	2	2	2	1	2	2	1	2	2	2	2	2	2	2
6	34	1	1	1	2	2	1	2	2	2	1	2	2	1	2	2	2	2	2	2	2
7	33	1	1	1	2	2	1	1	2	2	1	2	2	1	2	2	2	2	2	2	2
8	32	1	2	1	2	2	1	1	2	2	1	1	2	1	2	2	2	1	2	2	2
9	30	1	2	1	2	2	1	1	2	2	1	1	2	1	2	2	2	1	2	1	1
10	28	1	1	1	1	2	1	1	2	2	1	1	2	1	2	2	2	1	2	1	1
Total for lower		10	12	14	19	20	15	16	20	23	11	17	21	10	21	20	20	17	20	18	17
Diff. between		40	49	36	27	20	35	34	23	14	39	33	21	38	21	22	23	32	24	31	33
D.P. (Mean)		4.0	4.9	3.6	2.7	2.0	3.5	3.4	2.3	1.4	3.9	3.3	2.1	3.8	2.1	2.2	2.3	3.2	2.4	3.1	3.3

poor discriminatory power. Therefore, the researcher after completing the item discrimination analysis faces the problem as to which items should be included in the test and which are to be eliminated from the test. The test experts have not given any specific conventional guidelines for making decisions in such cases. Therefore, the researcher is always making an arbitrary decision in making the choice of the items. It is always better to take into account the mean scores of the items or the discriminatory power of the statement to accept or reject the specific statement or item.

From our discussion made earlier on the data given in Table 16.1, it is better to eliminate items 9, 12 and 14 and retain the rest of the items in the test. That is, the items whose mean scores or discriminatory powers are 2.2 and above may be selected to include in the test. The 50 per cent of the middle scores are not taken into account in our item discrimination analysis. Our assumption that only the extreme scores may provide the discrimination between the highs and lows clearly. The individuals in the middle ranges are more likely to score the items with response weights of 4, 3 and 2. By including the scores of these individuals in the item analysis, the discriminatory power, the mean scores of the items may become insignificant. This would invalidate the whole attitude scale.

The item discrimination analysis has certain advantages over other methods in determining the reliability of a test. This method helps the researcher to eliminate the items that are inconsistent with the rest of the items objectively. The inconsistency of the test items can be best spotted out only in this method and thereby the internal consistency and the reliability of the test can be improved. But some test users state that since there is no guidelines in the test development procedures with regard to elimination of items, the arbitrary elimination of items somewhat reduces the value of the item discrimination analysis. This need not be taken as serious limitation of this method as elimination is made objectively on the basis of experimental data.

17

VALIDITY OF MEASUREMENT

Introduction

Every researcher becomes very anxious whether the measuring instrument he has chosen or developed measures what it is supposed to measure. This anxiety of the researcher focuses on a very important characteristic of measurement, the validity of the measure. The measure of validity refers to the extent to which a test measures what we wanted to measure. It is to see whether the test actually measures what we intend to measure. For example, a tape is used to measure the length of a glass pane and its measurement should decide that the glass pane may fit in between the frames of the window in the room. Measurements of this type should always serve our purpose and it has done its purpose with proper validity. There is no difficulty with physical measurements, but in psychological measurements it is very difficult to establish the validity of a scale. It is a serious problem for every psychologist while constructing a measuring instrument that is valid and reliable. Sometimes, it may not measure for which it was constructed. Therefore, its validity and reliability has to be seen.

Validity is mainly based upon the amount of error in the measurement. There may be some mistake or difference in the test construction. Absolutely error free instrument to measure human characteristics is unthinkable. However, it is possible to postulate certain characteristics of the instrument which is applied to measure the desired trait.

What is Validity?

Validity has been defined by different authors in different ways. But it can be defined in short, as the absence of all kinds of errors. Validity reveals the merit of our measurement. In other words, it is said as the merit index of our measurement. Lindquist (1942) states that "the validity of a test may be defined as the accuracy with which it measures that what it is intended to measure or as the degree to which it approaches infallibility in measuring what it purports to measure." According to Cureton (1951) "validity of a test is an estimate of the correlation between the raw test scores and the true (that is perfectly reliable) criterion scores." A test is said to be valid if only its correlation coefficient is high. A test was constructed to measure the mechanical aptitude of a group and administered on them. Those who get high scores in the test must be successful in the jobs involving mechanical ability and those who get low scores must be unsuccessful in the job. After selecting the candidates for the job on the basis of the scores their scores are correlated with their performance. If the test scores correlate well with their performance then the test is valid. If not it is not valid. A test is valid only for the measurement of mechanical aptitude of the persons and it is not valid if it measures the clerical aptitude. If the test scores correlate well with the clerical jobs it is not mechanical aptitude test, but clerical aptitude test.

While correlating the test scores with performance or with a criterion score there will not be any absolute validity. We determine only the degree of validity. Some measures have high validity, others have moderate validity and some low or no validity. If we want to determine the rulers and the weights we use whether they are correct measures of things we can compare the instruments with other instruments which are accepted as standards. If the rulers and weights correspond to the accepted standard of measures then we accept and conclude that our instruments are valid.

The problem of validity is more complicated than the reliability as all the questions concerning a test, those related to validity are the most important and yet most difficult to answer. The collection of reliable information on test reliability is often expensive and also time consuming. The important question is: does the test measure what it supposed to measure? The answer can be given with simple yes or no answers. But how do we know that it measures what is supposed to measure. There are several ways of answering this question. The types validity of tests are classified by Thorndike and Hegan (1970) mainly in two types, one being the rational or logical analysis and the other being empirical or statistical analysis. The rational or logical analysis includes the contents and the topics of the test. When a rational or logical analysis refers to a specific concept it is called concept or content validity. The content validity and face validity are said to be based on rational analysis. Thorndike and Hegan (1970) have stated that it is also called rational validity. It has mainly to do with effective expression. Test items should have precise and concrete characteristics. The empirical and statistical analysis include construct validity, concurrent validity and predictive validity.

Types of Validity

Strictly speaking the types of validity are not having a clear cut distinction among them. Some of them are practically identical with one or two of them. But, however, it is possible to differentiate them as enough differences appear among them. According to Ebel (1966) all the types can be grouped under two major categories, such as: (1) primary or direct validity and (2) secondary or derived validity. The tests that have direct or primary validity represent faithfully the tasks included in them and proportionately the kinds of tasks provide operational definition of the trait or achievement to be studied. In the case of the tests that have derived or secondary validity the scores of the test yield correlation with the criteria scores. According to these definitions, the primary or direct types are: (1) validity by definition, (2) Content validity, (3) Curricular validity, (4) Intrinsic validity, (5) Face validity and the secondary or derived types are: (1) Empirical validity, (2) Concurrent validity, (3) Predictive validity, (4) Factorial validity and (5) Construct validity. The primary validity is evaluated and critically examined by the test constructor, whereas, the derived validity involves always correlating with some criterion test scores. Let us discuss some of the important validities more in details.

Content Validity

Content validity and face validity, according to Black and Champion (1976) are considered exclusively a logical type of validity. Measurement experts consider that content validity is crucial for achievement tests. It is always that an educational achievement test should possess content validity with more educational contents. The validity of an educational test is judged in terms of its relevance to the material of instruction rather than the ultimate objectives of instruction. Normally an educational achievement test is designed to measure the extent of attainment of the ultimate goals of instruction by the individuals in a particular subject or area. What is required of the test is, to what extent the test requires demonstration by the student of the achievements which make the objectives of the instruction in the particular subject or area. Therefore, the more fully and completely a test measures the attainment of these goals the greater the content validity is. In the construction of a verbal or quantitative test (question paper for examination) for S.S.L.C. and Higher Secondary Classes it is important that the tests have content validity.

In evaluating the content validity of a test it is important to observe whether the content of the test is suitable to the purpose for which it is designed. For example, the validity of an achievement test in geography should be assessed by determining whether the items dealt with only geographical subject-matter. While one is taking up an examination in geography finds on it a question or two on the rule of Emperor Asoka it can be reasonably claimed that the test or the item is not confining to the subject or area and may be stated that the test did not have the content validity. It is also important that the measuring instrument should be drawn from the universe of items. The universe of items may be consisted of all the facts or informations included in specific chapters of the textbook assigned to the students.

It is also observed that most of the educational achievement test users put more faith in qualitative comparisons of the behaviour called for by a test with the behavioural goals of instruction. But specialists in psychological testing do not trust this process. Because qualitative examination involves personal and subjective judgments. We are deceived by appearance. False relationships are accepted as true and true relationships are neglected as false. Black and Champion (1976) observe that one of the important problems associated with content validity is that it is subjectively determined. According to them, since content validity rests upon subjective judgment of the professionals to a large extent what one person states as high content validity may be stated by another as low content validity. The very divergent views held by both the teacher and the students towards the same examination is a clear indication of this issue. Therefore, many psychologists prefer a more objective, systematic and quantitative approach for determining the validity of educational achievement tests. Candland (1968) states that in preparing items for achievement test several common errors are possible. The test maker may fail to add questions which require specific answer or fail to make appropriate generalizations. A test has content validity if the items it contains represent adequately the variable the test is intended to measure.

Content validity mainly relies upon subjective interpretation of the individual researcher. It is not directly subjected to statistical verification. Therefore, there is too much chances for errors and mistakes in judgments. The validity established by this method is highly reliable. In spite of this, content validity has certain advantages. Content validity is directly applicable. It does not require more statistical knowledge. It is easy to apply and collect data and not time consuming like other validity methods.

Face Validity

Face validity is discussed under rational analysis by some test users. This is also called logical validity. Anastasi (1961) considers that it is essential in certain situations, though it is losing its importance in test validations. According to him face validity refers not to what a test necessarily measures, but to what it appears to measure. Guilford (1954) says that the 'test looks valid to those who are unsophisticated in test practices'. He further states that looking valid by itself is no guarantee for genuine validity of any form.

If a test is found to measure what is to be measured by another then it is said to have face validity. But the superficial appearance of some tests may create a confidence in test users and they may accept them and such acceptance is attributed to faith validity by some psychologists. If a test user thinks that a test is valid or relevant he may develop strong motivation for using it for his purpose. Cronbach (1966) says that a test which looks good for a particular purpose is said to have face validity. He further states that it is against scientific practice to adapt a test just because it appears reasonable and good looking. It was found that such tests have failed as predictors. According to Thorndike and Hegan (1970) face validity is "reasonableness or plausibility of test tasks, from the point of view of the persons being tested in terms of what he considers that the test is measuring". Sheridan (1979) states that face validity means that the validity of the items is determined by simply looking at them over and judging whether they test what they are supposed to test. All the college and school examination papers are validated in this way by the teachers. The meeting of the 'Question Paper Setting Board' of every university is held only for validating the examination question papers. Therefore, it is only by looking at the test and its contents that one decides that the test will be useful and serve the purpose for which he wants a test. This approach, looking at various aspects of the test and selecting it sometimes may disappoint the test user. But, however in certain situations this approach 'face validity' is made in choosing the test. In administrative, Army and Occupational services tests are chosen by face validity.

Many popular psychological measures used in magazines under the heading of "test yourself" for one quality or another were not validated empirically." They depend on what we call face validity. Face validity is better than nothing, but we must be very careful that it should not mislead us.

Construct Validity

The other approach of determining validity is the empirical and statistical analysis. This is the outcome of the relation the measuring instrument has with other fact corresponding to the test. The validity based upon empirical and statistical analysis is stated in three ways. They are: construct validity, concurrent validity and predictive validity. In construct validity the word 'construct' is an important concept. In construct validity the test to be validated measures a general construct such as aptitudes, intelligence, neurotic tendencies and so on. Candland (1966) says that there is no simple way of assessing construct validity. The concept, construct is not clearly defined, as such the problem of assessing validity becomes complicated. If the construct is well defined then construct validity is accurate. Thorndike and Hagen (1970) state that "construct validity refers to the accuracy with which the test describes an individual in terms of some psychological trait or construct". According to Payne (1967) a construct is referring to some granted attribute of people and it is supposed to reflect in the test performance. We make statements about attribute. It is putting together substances systematically in order to make a whole meaningful theoretical frame work. The test is intended to assess a particular trait and each statement in the test expresses something about the trait systematically. By this means the trait is fully stated or described in an organized manner. Every meaningful statement which expresses the attribute contribute something or a part to the constitution of the construct. For example, if cheerfulness is the attribute, the construct carries in every statement something associated with cheerfulness. Since a construct is an assumed attribute of people every test designed to assess some trait of a group becomes a construct. Thorndike and Hagen (1970) state that most names of traits refer to constructs. Thus a test designed to measure the trait 'cheerfulness' should really measure it. The name of every construct is a concept but every concept is not a construct. For example, cheerfulness, attitude, aptitude neurotic tendency are constructs. These refer to some aspect or characteristic of a person or group. Every one involves certain factors inherently. For example, cheerfulness is a trait of a person. He is having a happy disposition, in good spirits, pleasantly bright, ungrudging and enthusiastic. All these are attributes of cheerfulness. But there is no directly observable facts for which the word cheerfulness or anxiety stands. It is not possible for us to state anything more than the words as they are not directly part of our empirical world. We label such terms as constructs. Therefore, constructs are terms that have no direct empirical facts or referents.

But at the same time concepts are the terms that have direct empirical facts or referents. For example, 'book' is a concept. There is something empirical reality in this concept. The concept 'book' can indicate several objects. The properties of book represent the empirical reality. For example, the size of the book, pages, length, weight, thickness, subject-matter, type of print, colour and get up and so on. Table, chair, house, door, pen are also concepts and each one of these has direct empirical facts. Therefore, the term construct that we are using in this section to explain construct validity is different from the term concept.

The construct validity is pertaining exclusively to traits that are not directly assessed. In construct validity the test scores as a means of demonstrating construct validity of the instrument cannot be correlated with actual behaviour directly. The behaviour may be rated by the self or by outside agencies and these rating scores may be correlated with the test scores for validity.

Sometimes we expect high correlation between the scores of the instruments as evidence of construct validity and to show that the traits are the same and so they correlate highly. At other times we wish to show low correlation between the scores of the instruments to show that the traits are different and hence the low correlation. Further we also want to show the importance of the methods of measuring the traits. Because in construct validity it is suggested that atleast two traits should be measured by atleast two methods.

Construct validity according to Campell and Fiske (1959) is of two types. One type is convergent validity and the other is discriminant validity. The difference between these two types is operational in determining validity. The types of construct validity also states what does a test measure and also what it does not measure. When some new and universal trait measures are designed the tests must be interpreted and justified for validation and the construct validity should be made to accept. In such cases both the convergent validity as well as discriminant validity should be applied. The tests which are having too high correlations with other tests also may be invalidated as they are expected to differ from them with very low correlation or with no or negative correlation.

Convergent Validity

When a test or a measure of a trait correlates highly with some other kinds of instruments designed to measure the same trait is called *convergent validity.* The measurement of construct validity of convergent type is made by different methods. The rating method also may be used in the validation. Ratings may be made by the self (the test user) and other outside agencies. The scores of the test obtained from the sample and the rating scores may be correlated to determine the validity.

Further the scores of the performance tests and the printed tests also may be correlated to determine the validity. But we cannot say that any trait can be equally measured by all the methods. Because there are many traits that are not well assessed by all the methods as the behavioural signs of some traits are not observable equally by all methods.

Discriminant Validity

When a test or a measure of a trait yields very little or no correlation with tests or measures of other traits it is called *discriminant validity.* The method may be the same or some other methods, the correlation should be poor. But it is also possible that the tests or measures of traits by some other method may correlate higher than what it should be. This may occur because of the 'halo effect' as ratings are made by outside raters. The outside rater is inclined to rate the observed person similarly in all the traits and this makes the correlation higher. If we want to assess the discriminant validity and also to judge the contribution of the trait and the method, it is important that we must use more than one trait and more than one method in the validation process. The inter-correlations between the traits for each method may be presented in a multitrait-multimethod matrix. Each of several traits is measured by each of several methods. For example, if we measure three traits and by three methods, every one trait is measured by all the three methods. Hence, there are nine variables. Thus, we have $\frac{9(9-1)}{2} = 36$ inter-correlations in the matrix. In the multitrait-multimethod matrix we can find the reliability and validity of each trait.

Predictive Validity

We have stated elsewhere that there are two main categories of validity, direct and derived. Discussing specifically about derived validity, there are two kinds of derived validity. They are very much important for classroom test or achievement test constructors. They are predictive validity and concurrent validity. Correlation between earlier test scores and later measures of achievement indicates predictive validity.

Psychological tests are used for various purposes. One of the important purposes is predicting some specific future outcome. There are various situations in which such predictions are to be made with suitable tests. The scholastic aptitude tests are used to predict how well the high school students be successful in professional courses like engineering, medicine or agriculture or in liberal education in college. The vocational Counsellor and the personnel manager of some organisation and the industrial psychologists are all interested in making their prediction successful in the case of their clients when they are placed in some specific job. We expect that the person who has scored well in the mechanical aptitude test also should do well in machine operation. He should operate the machine without any damage and with high production and low personnel turnover. The person who scored high in clerical aptitude test should be an efficient clerk in an organisation.

For all these purposes the problem is choosing a correct criterion measure which can correlate with the job performance. Some test which is in practice and well accepted may be chosen as the criterion of success. The scores of the test may be correlated for estimating the validity. According to Thorndike and Hagen (1970) "predictive validity can be estimated by determining the correlation between the test scores and a suitable criterion measure of success on the job." Candland (1968) says that predictive validity is concerned with whether or not a test may be used to predict the future behaviour of the subject. Predictive validity may be estimated by comparing the test scores with one or more external variables (criteria) which can provide a direct measure of the behaviour to be assessed. The comparison may be in the form of a correlation between the test scores and the criterion measure. The test scores and the criterion data need not be collected at the same time. The criterion data may be collected at a later date than the test scores. But whether the test scores and the criterion data should be collected concurrently or at a later time depends upon the purpose. Predictive validity is the most useful type of validity. According to Magnusson (1967) Predictive validity is based on the measured association between what a test predicts the behaviour will be and the subsequent behaviour exhibited by an individual or a group of people. For example, we want to ascertain a particular religious groups' attitude towards family planning and birth control and the scores were obtained from them on the attitude scale. The group of people expressed their acceptance of the programme by scoring high on the scale. The same people's actual behaviour was observed in their home situation and rated and scores were assigned. These two scores, the expressed scores in the attitude scale and the actual behaviour rating scores were correlated and this will provide with an evidence of the predictive validity of the measure. Thus, predictive validity is estimated by the simple correlation between the scores obtained by the test and the behavioural scores obtained by rating in the home situation. A high correlation between the predictive scores (test scores) and the result (actual behaviour) means that the test appears to have predictive validity. It is the most reliable method for determining validity in most circumstances.

Generally predictive validity is determined by correlating test scores obtained at one time, that is at the time of employment with criterion measures—the job performance scores obtained at a later date (after placement). It is also called follow up method of test validation.

We are interested only in the degree to which the test scores correlate with some selected criterion measure. We judge the test in terms of its relationship to the criterion. The higher the correlation coefficient the better the test is. There are many types of jobs or occupations and it is very difficult to get a criterion measure for each job. The criterion measures available are only partial and that they measure only a part of success on the job. If suitable criterion measure is not available it has to be developed. Developing acceptable criterion measures in the field of tests and measurements involves a number of troublesome problems for the research worker. The problem in validation is always present in choosing the most satisfactory criterion of measures from among those that appear to be satisfactory. In the process of choosing criterion measures Thorndike and Hegan (1970) suggest few qualities that are desirable in a criterion measure. They are, in the order of importance (1) relevance, (2) freedom from bias, (3) reliability, and (4) availability.

Since the predictive validity is evaluated by empirical and statistical method it is also called empirical and statistical validity. Predictive validity is mostly applied to aptitude tests and achievement tests. Sometimes it is also applied to educational achievements and class-room tests.

Sometimes test users find it necessary to develop new tests though some tests are already in existence. This is, because there is every reason to believe that the new instrument is actually better than the test against which it is initially validated. The existing test may require the test administrator to read the instructions aloud, whereas, the new test may replace the procedure and may present printed instructions to the subjects which may be more valid. Alfred Binet had validated his first intelligence tests against the teacher's ratings of the pupil performance. Later on his tests became the standard test against which new intelligence tests were validated using Binet's tests as criterion measure.

Further, the existing test may be lengthy and may take an hour in administering the test. The test may be short and can be completed in half of an hour and thereby saving the time and effort. The new test is validated against the measurement which is called criterion (criterion measure). The predictive validity and the concurrent validity use the criterion measure. The difference between these two approaches is that *in concurrent validity the existing criterion is used whereas in predictive validity the criterion is available only after the test has been administered.* The behaviour is expressed only later on. The behaviour present or absent is the criterion score and thus is correlated with the scores obtained earlier on the test. The advantage of predictive validity is that it can demonstrate the association between predictive behaviour and the observed behaviour in numerical value. The technique adapted here really reveals prediction. In other methods this does not take place. The predictive validity has certain disadvantages also. It is not proved in predictive validity that the instrument really and absolutely measures the trait. Further it takes longer time in determining the validity.

Concurrent Validity

The predictive validity we have discussed here is not always practical to use as an evaluation procedure. It is not always possible to retest the subjects at a later date. What the experimenter or one who has devised a new test wants is to know how well the new test correlates with already well established tests. For example, an experimenter has constructed a test to predict the grades of engineering college students and wanted to assess its validity. For which, first he has to give the test to the first year students and obtained their test scores and correlate at the end of the fourth year when they finish their course of study with their grades. If the correlation is high and significant then the validity is also high and it is called predictive validity. What the test measured in the first year was found to be correct after four years in the grades of the students. This means those who scored high in the test at the beginning were able to get high grades at the end, after four years. This means the test correctly predicts what the students could perform at the end. This involves laborious task and longer time. That is why investigators prefer the concurrent validity to predictive validity. In concurrent validity procedure the problem is easily solved. In the case of predicting the grades of engineering college students the test is given to the senior students and their grades are compared with the test scores at the same time.

Concurrent validity is stated as derived validity. It refers to the relation of test scores to an accepted available contemporary criterion of performance on the variable which the test is intended to measure. Concurrent validity establishes the relationship between the test scores and the criterion measures at the same time. It differs from predictive validity with respect to criterion measures. In concurrent validity both the test scores and the criterion scores are obtained simultaneously whereas in predictive validity the criterion measures obtained only later on. Tests are developed in many situations and it is cumbersome to collect the information already in use. The problem in validation is whether the new tests agrees with the present source of information. If they disagree then the test measures something else. Therefore, validation is to be made again empirically. The test and the original procedure are to be applied on the same persons and the scores are compared. For example, the test scores obtained in a manual dexterity test are compared with judgments made by supervisors who rated the subjects' work in an industrial work which involves manual proficiency. This type of empirical comparison is called concurrent validity. In this process two sources of measurements or scores are obtained almost at the same time. A comparison of work productivity records with efficiency scores obtained at the same time may indicate whether the measures are valid according to concurrent validity method. The high correlation coefficient is the evidence of the validity measures and we can conclude that the concurrent validity is significant.

Generally concurrent validity is applied to educational or scholastic achievement tests by correlating the test scores obtained from the students with the marks obtained in school or the grades the classroom teachers assigned to the same students. One problem in this is that the teachers assigning grades to the students are not highly reliable. Another point is that the grades normally reflect the students achievements whereas, the test was not intended to measure. Aptitude tests are validated comparing the marks earned in school by the students.

In estimating concurrent validity the investigator should have some elementary statistical knowledge to apply the correlational method to compute the degree of association between predictive behaviour and observed behaviour. The investigator need not wait and watch for a long time to assess the respondents' behaviour. The respondents' behaviour can be assessed by using a similar test in existence and well established as a criterion test. In estimating validity of a new test any test already designed and well established to assess the particular variable or trait may be used as a criterion test. If only a similar test was not available then one can go for assessing the respondent's behaviour directly. Epstein and Shontz (1969) state that "since there is no Bureau of Standards for Psychology to provide criteria against which new tests may be validated, acceptable criteria are hard to find."

Apart from educational achievement and aptitude tests, concurrent validity now-a-days is applied in many of the areas since quite a number of reliable tests have been designed and well established and put to use. Concurrent validity is applied to personality and diagnostic tests, tests of adjustment problems, tests of neurotic tendencies, intelligence tests, Socio-economic status scales, attitude scales and so on. Therefore, any one test quite suitable may be chosen as a criterion test among the existing well established tests for the problem in question and both the new tests and the criterion test may be administered on the population selected for the study. The sample size in this case should not be less than 300. The two obtained scores of the tests may be correlated to estimate the concurrent validity of the new test or the test to be used to assess the particular trait.

Criterion-related Validity

In the preceding section we have discussed predictive validity and concurrent validity as derived validities. In both types the test scores are compared with the criterion. Both the types require the use of a criterion. The difference between them is essentially that the predictive validity employs a criterion at a later date, after administering the test, whereas concurrent validity employs one of the well established tests as a criterion. The American Psychological Association (1966) in its standard for educational and psychological tests and manuals discussed the predictive and concurrent validity under the same heading of criterion-related validity. There is some reason behind for combining these two validities. Practically it is not easy to distinguish between these two methods. The APA set forth only three basic types of validity of tests. These are criterion-related validity, construct validity and content validity. All these types of validities we have discussed here.

Intrinsic Validity

The intrinsic validity is stated as primary or direct validity, but involves experimental techniques other than correlation with the criterion. This validity has been discussed by Gulliksen (1950) in detail. Intrinsic validity provides objective and quantitative evidence that the test is measuring what it ought to measure. Guilford (1954) defined it as the "degree to which a test measures what it measures may be called intrinsic validity". This definition also may mean that "how well the obtained scores measure the test's true-score components." The intrinsic validity is stated in terms of square root of the proportion of true variance. It

is the square root of its reliability of the test. Guilford has defined in terms of the equation $r_{t\infty} = \sqrt{r_{tt}}$.
This statistics is called index of reliability.

$r_{t\infty}$ = Intrinsic validity.

$\sqrt{r_{tt}}$ = correlation coefficient between the two tests.

Since this validity depends upon reliability of the test the conditions that affect reliability also will affect the intrinsic validity. If the reliability of the test is accurately estimated then the intrinsic validity is also accurate. Present day researchers in India mainly apply this intrinsic validity to their tests before they conduct any investigation for their research work.

Relevant Validity

The relevant validity is somewhat different from all other validitics we have discussed here. It has certain rational basis in its definition and interpretation. Generally, validity is concerned with the degree of agreement with the existing reality. In other words it deals with what a test measures. There are certain relationships between reliability and validity of test scores. This we have stated in the intrinsic validity. Here it becomes necessary to extent the rational basis given to reliability for giving reasons for the facts of validity. The degree of validity is indicated by some correlation coefficient. Validity is often described in terms of a correlation coefficient between the test scores and some measure of performance in some life situation. This correlation coefficient should be rationally explained. The multiple factor theory is a very good approach to explain this phenomenon. According to this theory there are two components in the obtained scores called true variance and error variance. The true variance can be further broken into two components called common factor variance and possible specific variance. The common factor variances may also be shared by other tests. The specific variance component is supposed to be unique to the test. Hence, the true variance consists of common factor variance and specific variance. If the proportion of specific variance (S_x^2) is deducted from true variance (r_{tt}) it is called communality (h_x^2). The equation is: $h_x^2 = r_{tt} - S_x^2$. Hence, communality according to Guilford (1954) is the proportion of common factor variance in the scores. Communality is symbolized by h^2. Guilford states that "the degree to which a test measures factors that are common to other measures may be called its relevant validity". The relevant validity is indicated by the square root of the communality (h^2) of the test, or the h.

The relevant validity, therefore, is similar to intrinsic validity. The only difference between the relevant validity and intrinsic validity is that the specific variance is removed in the estimation of the relevant validity. But in the case of intrinsic validity the specific variance is permitted to enter in the estimation. But, if the specific variance is zero then the validities are identical. The value of relevant validity is indicated by the extensity of communality (h). If a test's correlation coefficient with any other measure is higher, then the coefficient of h also is higher.

The important components of validity are reliability and relevance. According to Ebel (1966) "a test which measures with a high degrees of accuracy what it is intended to measure is highly valid because it is both highly relevant and highly reliable". Relevance is a matter of logical analysis and expert judgment for all the tests of educational achievement. Huddleson (1956) states that relevance ordinarily cannot be measured statistically. Relevance must be built into the test. Ebel (1966) says that this kind of validity is based on relevance and reliability and this ought to have applied to tests of educational achievement. It is primary, direct, intrinsic validity and sometimes referred as content validity and curricular validity. It is important to look at the test itself and at the rationale and specifications for the test to determine the level of this validity a test possesses.

18

NORMS AND TEST SCALES FOR MEASUREMENT

Introduction

Psychological studies were very much nourished mostly by the testing procedures of the first and second quarters of twentieth century innovators. Quite a number of tests and instruments were designed and developed during this period for the study of individuals' behaviour for evolving a class of behaviour and to generalise the modes of behaviour or traits. Psychological testing itself has become a separate field of study. In the universities it has been included in the curriculum at postgraduate levels. Psychological testing has been direct and also indirect. This depends upon the nature of the traits included in the study for assessment. In direct assessment psychologists put more emphasis on observable behaviour of the individuals. How well a person can perform the task? He may be proficient in more than one task. Therefore a person's skills in more than one area also may be assessed directly. The individual's performances in the test may reveal what actually he can do on the job. This kind of testings has been classified as the measurement of achievement. Thus, achievement tests of various kinds have been developed by professional psychologists and academic psychologists. In selection and placement in business and industrial organizations and in the field of education and guidance achievement tests have been used to assess the skill. Therefore, psychological testing has become not only a field of study academically but also a professional field. Research papers have been published enormously. Eminent authors like Anastasi (1961), Thorndike and Hagen (1970), Cronbach (1966), Lyman (1963) and others have published valuable books on Psychological testing.

What is Psychological Testing

Therefore, what is psychological testing? It is defined in terms of the use of the psychological tests. According to Moskowitz and Orgel (1969): "the construction and use of scales for the measurement of individual differences is called psychological testing". It is the application of the tests to assess the individual's behaviour and comparing them for evolving a norm or standard. According to Anastasi (1961) a psychological test is an instrument for obtaining an objective and standardised sample of behaviour. Therefore, psychological tests are not only intended to assess the individual's achievement but also to assess various other human potential abilities, intelligence, aptitudes, interests, attitudes and prejudices. The achievement tests measure what one can really perform at the time of testing. Whereas, the aptitude tests and other tests measure one's capacity or his potential abilities. This is inferred, not directly observed. For example, aptitudes, intelligence, interests, attitudes and prejudices are inferred from test results the individual's behaviour as to how he would behave in future.

There are certain common characteristics for all psychological tests. Every test is supposed to elicit responses of what is being measured and which should be the class of behaviour of the group. The performance of a test in this context is like an experiment controls and manipulates the variables. Every item of a test is like a stimulus eliciting a response from an individual. Just like the experimenter interpret responses, recorded by him from the respondent at the time of testing, as dependent variable, the test user also controls certain variables that might influence the respondent's responses and interprets the data. In both the situations other outside variables are controlled from contaminating the results of the experiment or investigation.

In psychological testing the construction of test is one of the important aspects. The test user may not get readily a test for the purpose he wants. Every test is designed and developed for a particular purpose or to measure a specific trait or variable. Even if one such was available it is important that it should be again revalidated before being administered in a population. If a suitable test is not available we should design and develop one. The test items should be carefully selected and tested in a sample and the sample scores should reveal the behaviour being measured. Every item should characterise some aspect of the problem to be studied. It should be empirically verified.

Every item should express one idea about the problem or aspect of the problem and should reveal the individual's reaction to the particular aspect. Thus an item is like stimulus reveals what an individual feels about the idea expressed in the item. The individual's reaction may be positive or negative or even neutral. The item reveals the individual's behaviour in this way. For example, if one wants to measure dominance-submissive he must use words and phrases that should reflect some characteristics of dominance or submissiveness. The very utterance of the word should demonstrate the characteristics. The items carrying the words relating to dominance and submissiveness should be verified experimentally whether they are identifying the characteristics and elicit the behaviour to be measured. It is better to verify this by administering the items in a population more than once with varied samples on different occasions. Because one sample to measure the behaviour in a population is not sufficient. It may not be the representative of total behaviour of dominance-submissiveness or the characteristics in which we are interested since it may be sometimes influenced by extraneous variables.

Further, every characteristic of the variable included in the item should have equal chance of being represented in the test proportionately. For example, if dominance is described using twenty different type of words, each word indicating one aspect of dominance then these twenty word should find place in the items, one aspect in each item. These items should have equal representation in the test, so that every aspect of dominance is included in the test and there will be proportionate representation of the words expressing the trait. It is also possible to express the characteristics of a trait in too many words and hence we can have too many items. Similarly, the trait may be described with very few characteristics and hence very few items. This type of approach of test construction may result in having either a too lengthy test or too short test. Both types are defective. If a test is too lengthy, then the respondents may find it difficult to answer all the items and also find it unwieldy and loose interest in answering the items. After answering some items carefully they may start answering the items hurriedly without even reading them properly. This type of answers may lead to asymmetrical distribution of scores. It will have skewed distribution. Similarly, if there are very few items in the test, it will not be representing the whole trait. There will be inadequate representation of the trait in the test. Therefore, the number of items should be optimum enough to include all aspects of the trait and also the participants should find interest in answering such items. The test designer should empirically verify more than once the length of the test he is using for his research. He should be satisfied with the length of the test by a statistical technique and also create confidence in others.

Test Scores

Psychological testing is made with a definite purpose of getting a quantitative assessment of the behaviour of individuals or group of people. Quantification is made by converting the responses of the individuals into numerical values. These numerical values are called scores. The scoring is made according to the procedure designed for the test. It is not done subjectively according to the judgment or view of the individual scorer or examiner. It is done objectively. Generally, good tests have well defined rules for administering, scoring, checking the correct responses and assigning numerical values to the responses. Certain tests have strict instructions in administering, in taking up the tests and in getting the correct scores for the responses. There is no one single way of scoring the test responses. There are several ways of scoring. This depends upon the type of behaviour we study and the purpose for which the study is made.

It is true that some tests are easily scored and quantified. For example, the speed tests or some intelligence tests are scored easily as they are time bound tests. There are tests answered with yes or no, true or false, agree or disagree responses. These are all 'either' or 'type' of responses. Certain tests require more time to answer. Response categories for some tests may vary from 3 to 7 or 9 or 11. For example, tests developed according to Likert scale technique have five responses and Thurstone's equal appearing interval method of scaling has eleven responses.

The participants while answering such test items will have to take more time for deciding one correct response out of five or seven. There are tests taking still more time for answering, for example, Rorschach Inkblots test, Thematic Apperception Test. In Inkblots test the respondent is asked to tell what he sees in ten inkblots. The inkblots are having irregular shape so that innumerable interpretations can be made. They are designed mainly to arouse emotional responses from the participants. This is used in clinical situations. The interpretation of the scores can be done objectively. There are about a dozen major scores and all of them fall into three major categories, location, determinants and content. The Thematic Apperception Test on the other hand, requires the participant to tell a story from the picture. He has to tell what is happening there, what has led up to the scene and what will be the outcome of this. The aim of this test is to see to what extent the participant projects himself in the scene and his identification with a character and the type of role he takes in the scene. The TAT consists of 20 pictures and there are separate pictures for males and females. The story told by the participant may indicate the attitudes he has taken in each situation. The scoring can be done objectively. The TAT mainly reveals the individual's personality traits. In both the tests the responses elicit the individual's emotional reactions and personality traits. But both the tests may take longer time and the scoring is not simple. The scoring and interpretation can be done by only experienced test users.

Norms

Every test is having a scoring procedure and every test score has to be interpreted meaningfully. A score obtained from an individual in a test is called a raw score. This raw score is not used as it is for interpretation. Because a raw score as it is has no meaning. For example, a score of 75 obtained in a psychology test or a score of 90 in an attitude scale does not convey any meaning. If the score must be meaningful it has to be compared with some standard score of test. In psychological measurements standards are based on average or the mean of the test performance of a group of people. The individual's score is compared with the mean score of the group or sample. The score of 75 obtained by a person in a psychology test becomes meaningful if we come to know that the average score of the group in which he is also a participant is 70. This means that individual scored higher than the mean. The mean score of the group in the attitude scale is 105 whereas, the individual obtained only 90. His score is less than the group mean. The average or mean score is a standard score against which the individual's score is compared. This type of standards are obtained by empirical method. Norms can be developed on the basis of the nature of the test performance by a group of people.

In the test standardization process the development of norm is one of the important aspects. Having obtained a confidence in a test by administering on a representative sample the test designer choose randomly a larger sample from a specific population and administer the test once again. The mean score is computed as a standard and any individual's score is compared with it. Norms for psychological tests are assigned in various ways. Norms also are depending upon the independent variables of the sample. The norms mainly serve as the comparison scores.

Norms, according to Thorndike and Hagen (1970), "constitute frames of reference for the interpretation of test scores. They are based on the actual performance of groups of specified types." The scores of individuals are compared in terms of age, grade, occupational status and so on as norms are developed on the basis of these variables.

Norms are better ways of expressing the test results in a meaningful way by applying statistical operations. The raw scores and the percentage scores are in crude forms. Therefore, they need to be transformed into some common scale for comparisons. The norms also have certain limitations. For example, the centile values represent only the ranks.

Age Norms

Since age is an important independent variable, tests like scholastic aptitude tests need age norms for comparing the same age group of different cultures, rural-urban or different socio-economic status groups. This is because with age there may be progressive change in the trait. Therefore, the age norm becomes a necessary standard. Thorndike and Hagen (1970) state that age norms as a system of test norms based on age equivalents. It is the average value of the trait of the persons of the particular age. Age norms are widely prevalent in achievement and aptitude tests. According to Ebel (1966) an age norm is the average score on an aptitude or achievement test for pupils of a particular age group. Age norms are normally presented in tables, the average scores of students of different age groups serially.

Age norms for children studying in elementary schools can be determined. Right from the age of 4 or 5 the average age can be determined up to the age of 14 years. Since the height of the children increases with age there can be a norm for every age in height. If his height is more than the age norm he grows faster. If he is less than the age norm he grows shorter. In the same way age norms for every trait can be determined and the individual may be compared. But age norms are appropriate for traits that have normal growth or traits that make progressive changes. Traits that do not have continuous change on the basis of the age cannot be stated in terms of norms. For example, the case of visual acuity or sense of hearing norms cannot be determined. Age norms may be based on large samples of pupils throughout the nation.

Test designers face practical problems in determining age norms for certain traits. First of all it is very difficult to get a truly representative sample of the pupils of the same age in large number and bring them together for testing. One age group is not sufficient for estimating age norms. Atleast the whole of elementary school grade pupils ranging from 5 to 14 years should be included in the age norms. This is not an easy task for the test designer. However, age norms can be determined only for the elementary school grade students. As the students grow into adolescents or adulthood age norms do not bear any meaning in terms of performance. Age norms are acceptable only in certain characteristics such as height, weight, intelligence and certain physical abilities. Since progressive changes are taking place every day, what is estimated in the first month of a year may be different in the second month. Therefore, if age norms are to be estimated it should be for every month. The school grade achievement may be varying month to month. This is what Thorndike (1970) has observed in using his Large-Thorndike Intelligence Test.

Grade Norms

Grade norms mainly refer to the achievements the pupils make in schools grades. It is the average achievement of a particular grade pupils. A norm is estimated for a grade group. According to Ebel (1966) "a grade norm is the mean or median achievement of pupils in a particular school grade on a given standardized test. Grade norm tables present these mean scores for several adjacent grades." Test designers when used standardized tests they have also transformed the raw scores into grade equivalents or grade norms.

Grade norms are estimated on the basis of large representative samples of the pupils throughout the nation just like the age norms. A test is administered to the representative samples in each of a series of school grades throughout the nation and the average is determined for each grade. For example, if the grade norm is 7 for a pupil means that his score is equivalent to the average score the sixth grader makes in the beginning. If it is 7.5, his score is equivalent to the average performance at the middle of the grade and so on. It is also true that pupils in some parts of the country claim that they are superior to others. This is because of the differences in pupils' ability, quality of teachers and facilities available for the pupils for their studies.

The interpretation of grade norms and the individual score in a test should be done carefully. A fourth grade pupils from educationally advanced family was found to be very bright. A standardized arithmetic test is administered and he obtained the score which is equivalent to the grade 6.5. This should not be taken for granted that the pupil has mastered the arithmetic taught to sixth grade. The pupil has obtained the score equivalent to the score obtained by an average pupil in the middle of the sixth grade. This score is the maximum possible higher score that a fourth grader can obtain in the arithmetic test and this does not qualify him to go straightaway to seventh grade.

According to Cronbach (1966) grade norms are not standards and it should not be mistaken that all pupils in the ninth grade should reach ninth grade norm. He says that 50 per cent of the pupils fall below the norm and grade norms are based on unequal and artificial units of measurement. It gives an impression that if a pupil of 5th grade was much superior if he obtained eighth grade level in arithmetic. Therefore, Cronbach (1966) is of the view that grade norms should be replaced by percentile scores based on single grade groups of a specified school as grade norms may lead teachers and parents to wrong conclusions.

Percentile

The concept percentile is associated with several values. Percentile is closely associated with quartiles. Here the *P*th percentile is a point below which *P* per cent of cases fall. Thus, the median is 50th percentile, the third quartile is 75th percentile and the first quartile is 25th percentile. The 10th, 20th, 30th..... 90th percentiles are called deciles. A decile is one of nine actual values of a variable dividing its distribution into ten groups with equal frequencies. But the location of zeroth and 100th percentiles are difficult to locate as they are influenced by chance. Hence, it is suggested to drop these two percentiles. Further, percentiles are associated with the cumulative frequency distributions.

Percentiles are the values by which the psychological and educational test scores of the individuals are reported. The raw score ordinarily does not give any meaningful interpretation. But if it is stated that an individual's score is near 75th percentile it implies that 25 per cent of his group pupil scored higher than him. This gives the idea of the location of the individual in the reference group.

Apart from the concept percentiles, test designers use also the concepts percentile scores, percentile ranks, percentile norms, and percentile band. These concepts often may cause confusion to the researchers. Even with regard to percentiles, Guilford (1987) states that it is consistent to call for "centiles" and not "percentiles" though it seems to persist in popular use. Whatever, the defference of opinion persists with regard to the usage of the concepts, it is better that the student should know what these concepts refer to. Though we may use the concept percentile in different froms the subject-matters of these concepts do not vary too much.

Percentile Scores and Percentile Ranks

In a testing situation individuals are compared on the basis of their performance or scores they obtained in the test. The easiest way of comparing individuals is to arrange the scores from highest to lowest and to rank them. In this arrangement it is possible to state where a particular pupil stands. It may be also indicated by numerical values as 2/60, 5/60 and so on. It means that the two particular pupils stand in the second and fifth positions or ranks out of 60. Ranking depends upon the number of persons in the group in which the particular person also is a member.

It is better to change the ranks into percentile scores to avoid problems involved in change of the size of the group. The percentile score is the rank stated in terms of percentage. The percentile score of a person can tell us how many of the persons in the group are above him and how many are below him. Suppose, for example in a class of 60 pupils there were 15 above Mr. X and 44 pupil below him. In ranking the pupils on the basis of the scores X is in the 45th place. By dividing case X arbitrarily between two groups we get 15½ and 44½, which means that 15½ cases are above him and 44½ cases are below him. Now work out the percentage for 44½. It is $(44\frac{1}{2} \times 100) \div 60 = 74$. The pupil X's percentile score is 74. By computing the percentile scores in this way the pupil who falls exactly in the middle is at the 50th percentile. This is at the centre of the distribution and, therefore, the 50th percentile is the median and median is the proper measure of central tendency in examining the percentile scores. Here, as stated earlier the ranks are changed into percentile scores. Therefore, it is also called percentile ranks and since percentile is in other words stated as centile it is also called centile ranks.

Percentile Norms

In age norms the average scores of the pupils of a particular age group are considered. In grade norms the mean scores of the pupils in a particular school grade on a given standardized test are considered. In both the estimations comparisons are to be made. But in percentile norms comparison of the individual score is made and interpreted in terms of his standing in a particular group of which he is a member. Percentile norms, according to Thorndike and Hagen (1970) is a system of norms based on percentiles within a specific reference group. In other words it is comparing an individual pupil in terms of his achievement to his own age group or to his grade group of which he is also a member.

Percentile norms require large number of tests to be administered on a bigger representative sample and norms should be computed for each test for the group. Also better to compute norms for different age groups and grade groups. So that a 7 years old is evaluated in terms of 7 years old norms, a 5th grade pupil is evaluated in terms of 5th grade norms. In the same way it is also possible to have norms for occupational groups and an applicant for a post of clerk may be evaluated in terms of clerk norms.

Percentile Band

Percentile Band is another way of evaluating the score in a meaningful way. Though it is not used by the researchers in their treatment of the data it is better to know what it is. Percentile band, according to Thorndike and Hagen (1970) is a range of percentile values within which the true percentile score of an individual may be expected to lie. Percentile band supposed to extend one standard error of measurement above or below the percentile rank. It is to make the test user to be aware of the accuracy of his test scores and attach importance to the precision. Percentile band is like a standard deviation dealing with dispersion of the score. For a particular raw score of an individual where his percentile score is supposed to fall is indicated in the percentile band. Percentile norm indicates where the individual stands in the group in which he is also a member.

Standard Scores

The percentile scores we have discussed in the preceding section do not convey the same meaning throughout the range as the units of a system are not uniform. Therefore, we look for a derived score from the raw score and the particular kind of derived score that plays an important role in statistical analysis is the standard score which is symbolized by z. The standard score is normally defined in terms of the formula:

$$z = \frac{X - \overline{X}}{S} = \frac{x}{S} \quad \text{or} = \frac{\text{Raw score} - \text{Mean score}}{\text{Standard deviation}}$$

Where X = the raw score

$\overline{X}$ = the mean score

S = the standard deviation of the distribution.

This is one form of transforming the raw scores to a new set of scores which are comparable. Thus a standard score has a known mean score and a known standard deviation. In this new set of scores it is possible to compare the scores of two psychological tests of an individual. But this is not possible with raw scores. This is one main advantage of standard scores. The standard scores according to the formula given above may involve both positive and negative values and also decimal scores. To have positive, negative and decimals in a small equation looks very uncomfortable for computation. Therefore, a further transformation is made in such a way that the uncomfortable position is removed. For example, if the raw score is smaller than the mean it will have negative value. Or the mean or the standard deviation may be having decimal values. The standard deviation has its meaning for most purposes only when two or more standard deviations are being compared. For example, if an intelligence test is conduced among three class of students one class had a standard deviation of 8, for the I.Qs, another class had 16, and the third class had 20. This means that the second and the third groups were having more dispersions in the individual scores and the two groups were more heterogeneous from the first group with regard to the intelligence quotient. Therefore, the reason for discussing the standard deviation in standard scores is that the standard deviation is the basic to the standard scores. This is referred to as *z* score. The *z* score shows how many standard deviation units score is from the mean. It can be on either side of the mean. For example, in a distribution the mean score is 40 and the standard deviation is 8. Then the *z* score corresponding to the raw score 48 is:

$$z = \frac{48 - 40}{8} = \frac{8}{8} = 1$$

This shows that the raw score of 48 is one standard deviation above the mean. Suppose if the raw score is 28, then the *z* score is

$$z = \frac{28 - 40}{8} = \frac{-12}{8} = -1.5$$

The *z* score is –1.5. It means that the raw score is 1.5 standard deviation below the mean.

Let us assume that in a frequency distribution the group mean is 65.5 and the standard deviation 16.25. The standard score for a raw score of 82 is computed as follows:

$$z = \frac{82 - 65.5}{16.25} = \frac{16.5}{16.25} = 1.015 = 1.02$$

It can be observed that the raw score of 48 in the first distribution is comparable to the raw score (82) of the second distribution with regard to the *z* scores. The *z* score is both distributions are almost the same (1.0 and 1.02). The relative distances above the mean in both the cases are same. Thus in this method an individual's performance in one test may be accurately compared with his performance in another test. It is also possible to combine the scores from several tests and comparisons may be made.

The use of percentile ranks in comparing the scores of two or more tests appears to be not clear and obscures the differences. Therefore, in combining the scores from several tests it is less accurate. In single test scores there will not be any problem in comparing the pupil and the teacher is safe in using the percentile ranks. However, the *z* score is more safe and accurate in comparing pupils' performances.

T-Scores

But in some situations the *z* score also is found to be undesirable, because it involves using negative scores and decimal points. To avoid this a modified form of the standard score is suggested. The formula suggested for this is to multiply the *z* score by a constant in order to eliminate the decimals and to round up the nearest numbers and adding another constant which should be large enough to eliminate the negative values. This procedure is called *T*-scaling and the resulting score is known as *T*-score. The formula to obtain the *T*-score is:

$$T = 10z + 50$$

The *T*-score is a normalized standard score.

This is another way of using the percentile scores for comparisons. The fact is that the mean of any distribution of standard scores will be always equal to zero and the standard deviation will be always equal to 1. The standard score is a derived score from the raw score and the advantage of it is that one standard score may be compared directly with another standard score of another distribution. The *z* score system converts the raw score to a value with *M* (Mean) = 0 and the *SD* (Standard Deviation) = 1. This sort of conversion is not satisfactory for our purpose because there may be negative values and decimals. Therefore to have a more convenient transformation is to have two scales as follows:

$$z\,50 = 10\,z + 50$$

$$z\,500 = 100\,z + 500$$

z 50 system is to have the mean as 50 and the *SD* as 10 and *z* 500 system is to have the mean 500 and the *SD* as 100.

By this system any *z* score may be multiplied by the desired *SD* and added by the desired mean.

Illustration with Example: Computation of Percentile Scores/Ranks

The derivation of percentile score or percentile rank, the standard score and the *T*-score may be illustrated with example here. In an achievement test 17 students from three schools participated and the following scores were obtained by them.

34, 37, 36, 35, 41, 43, 42, 44, 38, 40, 39, 48, 45, 47, 46, 50, 49

The first step is to rank the scores from highest to lowest. The highest is the first place and the lowest is the last place. The pupil in the middle of the group is at the 50th percentile. The 50th percentile is the median. If a pupil stands third out of 17 his position is relative to others. This ranking depends upon the number of persons in the group. If we are to compare the individuals in more than one test, then there is a problem of change of position. To avoid this problem ranks are changed to percentile scores. The percentile score in other words may be stated as percentile rank or centile ranks. Now let us change the ranks to percentile ranks or scores. There are two ways of changing the raw scores to percentile scores or ranks. A person's percentile score shows how many are above him and how many are below him in the group. In our example there are 17 persons. In the ranking Mr. X was placed in 7th place. This means that there are 6 pupils above him and 10 pupils are below him. But while dividing the group into two we will have to divide the position of X. In such case the rank of X, 7 is divided as 6½ and 10½. This means 6½ cases above X and 10½ cases below X. What is the percentile score for 10½? It is, 10½ × 100 ÷ 17 = 62. The percentile score of X is 62. First we must start at the lowest point 1, the 17th rank. It is in between 0.5 and 1.5. Therefore, for the last rank the percentile score is $0.5 \times 100 \div 17 = 2.94$. This must be rounded as 3. Similarly, the 16th rank is between 1.5 and 2.5. The percentile score is $1.5 \times 100 \div 17 = 8.82$ and it is rounded as 9. All other scores are computed in the same manner.

The other way of determining the percentile score not by dividing the common rank into two halves, but by tallying scores two times (Table 18.1). It may be called double tally method. Let us illustrate both the methods of determining the percentile scores or ranks for the 17 scores.

Table 18.1 : Estimating the Percentile scores by the method of arbitrarily dividing the cases

Raw Scores	*Ranks*	*Cumulative position*	*Percentile score*	
50	1	17	97	
49	2	16	91	
48	3	15	85	
47	4	14	79	
46	5	13	74	
45	6	12	68	
44	7	11	62	Model shown
43	8	10	56	
42	9	9	50	
41	10	8	44	
40	11	7	38	
39	12	6	32	
38	13	5	26	
37	14	4	21	
36	15	3	15	
35	16	2	9	
34	17	1	3	

In the double tallying method also the scores are arranged in rank order, the highest at the top and the lowest at the bottom. (Table 18.2). In double tallying method we are avoiding the awkward position of splitting the mid rank to show that half of them are falling above the mid point and half of them below the mid point. In double tallying method it is important to continue the score one step above the highest and one step below the lowest.

Table 18.2 : Showing the estimation of percentile scores by double tally method

1	2	3	4	5
Raw Scores	*Tally of scores*	*Cumulative frequency*	*Percentage of frequency*	*Percentile Rank*
51	1	34	100.0	100
50	1 1	33	97.1	97
49	1 1	31	91.1	91
48	1 1	29	85.3	85
47	1 1	27	79.4	79
46	1 1	25	73.5	74
45	1 1	23	67.6	68
44	1 1	21	61.8	62
43	1 1	19	55.8	56
42	1 1	17	50.0	50
41	1 1	15	44.1	44
40	1 1	13	38.2	38
39	1 1	11	32.3	32
38	1 1	9	26.4	26
37	1 1	7	20.5	21
36	1 1	5	14.7	15
35	1 1	3	8.8	9
34	1	1	2.9	3
33		0	0.0	0

Every rank is tallied twice and entered in the 2nd column, one tally just above and one just below the mid point of the score. The third column is the cumulative frequency of the tallies. The cumulative frequency is worked out from the mid point of the score interval. For the score interval 33 it is 0. For the mid point of 34 it is 1 and for the mid point of 35 it is 3. To the midpoint of 36 it is 5 and so on. At the mid point of 50 the cumulative frequency is 33 and at the midpoint of the last interval 51 it is 34.

In the fourth step the cumulative frequencies are converted into percentages. The formula is to divide the frequencies into 34 and multiplying it by 100. The bottom frequency 1 against the just below the score interval 34 is divided by 34 and multiplied by 100 $(1 \div 34) \times 100 = 2.9$. This value is entered against 34 in column 4. The other cumulative frequencies are also converted into percentages and entered against midpoint of the respective score intervals.

The last step is to round up these percentage values to the nearest whole number. For example, 2.9 is rounded as 3. 14.7 is rounded as 15 and 32.3 is rounded as 32 and so on. The rounded whole number values given in column 5 are the percentile scores or percentile ranks.

Percentile ranks are estimated on the basis of relation between a particular pupil's score and the scores of other pupils in the group. The percentile ranks must range from 0 or from near 0 to 100 or near 100.

Computation of *z* Scores

Let us estimate the *z* score using the same data. The important values required for computing the *z* score are the means and standard deviations. The formula is $\frac{X-\overline{X}}{SD} = \frac{\text{Raw scores} - \text{Mean scores}}{\text{Standard deviation}}$. The scores may be reproduced here for illustration. Raw scores: 34, 37, 36, 35, 41, 43, 42, 44, 38, 40, 39, 48, 45, 47, 46, 50, 49

$$\text{Mean} = \frac{X}{N}\ \frac{714}{17} = 42.\ SD = \sqrt{\frac{\Sigma X^2}{N}}$$

X	d	d^2
35	–7	49
37	–5	25
36	–6	36
34	–8	64
41	–1	1
43	1	1
42	0	0
44	2	4
38	–4	16

X	d	d^2
40	–2	4
39	–3	9
48	6	36
45	3	9
47	5	25
46	4	16
50	8	64
49	7	49

$$SD = \sqrt{\frac{\Sigma X^2}{N}} = \sqrt{\frac{408}{17}} - \sqrt{24} - 4.89$$

Mean = 42, SD = 4.89

The z score for the first score, 34 is: $\frac{34-42}{4.89} = \frac{-8}{4.89} = -1.63$

In the same way for all the 17 scores the z scores were computed and given in Table 18.3.

The z score for 34 is: –1.63, for 35 = –1.43, 36 = –1.23, 37 = –1.02, 38 = –0.82, 39 = –0.61, 40 = –0.41, 41 = –0.20, 42 = 0, 43, 0.20, 44 = 0.41, 45 = 0.61, 46 = 0.82, 47 = 1.02, 48 = 1.23, 49 = 1.43, 50 = 1.63.

These z scores consisted of both minus scores and plus scores. We can interpret the z scores in terms of standard deviation. For example the first pupil's z score is –1.63. We know immediately that his score is 1.63 standard deviation below the raw score mean. In the case of the pupil whose raw score is 50 the z score is +1.63. This means that his score is 1.63 standard deviation above the raw score mean. If it is minus it is below the mean and if it is plus it is above the mean. These derived scores based upon the standard deviation are called standard scores. Therefore, a z score is a standard score. According to the statistical principle in a complete distribution of z scores the mean is zero and the standard deviation is one.

Computation of *T*-Scores

These values (standard scores) are not very much satisfactory and useful for our purpose as the values are with negative and positive signs and in decimal points. Therefore the z scores themselves are to be transformed into more convenient scales. There are two such scales suggested. These scales have two constants. They are:

$$z\,50 = 10_z + 50$$

$$z\,500 = 100_z + 500$$

In the scale $z50$ the mean is 50 and the standard deviation is 10. In the scale $z500$ the mean is 500 and the standard deviation is 100. We can use the scale whichever is convenient for us. When we apply this scale on the z score it is called *normalized standard score scale* and indicated by the letter T and is designated as T-score. The first scale $z_{50} = 10z + 50$ may be applied to the z scores to convert them to T-scores. The formula is: $T = 10(z) + 50$

Therefore, the T-score for 34 is $= 10(-1.63) + 50 = 33.7$

The value 33.7 can be rounded to the nearest whole number and therefore, it is 34. All the other z scores are converted to T-scores and presented in Table 18.3.

Table 18.3 : Showing the raw scores, percentile scores/ranks, z-scores and T-scores on achievement test for 17 pupils

Rank for Raw Score	*Raw scores*	*Percentile scores—Deviation Method*	*Percentile scores—Double Tally Method*	*z-scores*	*T-scores*
	51		100		
1	50	97	97	1.63	66
2	49	91	91	1.43	64
3	48	85	85	1.23	62
4	47	79	79	1.02	60
5	46	74	74	0.82	58
6	45	68	68	0.61	56
7	44	62	62	0.41	54
8	43	56	56	0.20	52
9	42	50	50	0.00	50
10	41	44	44	−0.20	48
11	40	38	38	−0.41	46
12	39	32	32	−0.61	44
13	38	26	26	−0.82	42
14	37	21	21	−1.02	40
15	36	15	15	−1.23	38
16	35	9	9	−1.43	36
17	34	3	3	−1.63	34
	33		0	0.00	

The percentile scores/ranks are estimated by the method of dividing the midpoint into two halves and also by double tally method. The percentile scores/ranks obtained by both the methods are the same. In the double tally method since we extend the interval scores at the bottom and at the top the percentile score starts at 0 and ends at 100.

The z-scores are with decimals and negative and positive values. The negative and positive values when added become zero. The midpoint of the z-score is zero, whereas in all other norms it is 50. The T-score also is 50 since the z-score becomes zero when it is multiplied by the constant $10(z \times 10 = 0)$.

In a group of 17, a pupil who has obtained a score of 48 in the achievement test is placed in the 85th percentile rank and a pupil who has obtained a score of 40 is placed in 38th percentile rank. The advantage of percentile rank is that it differentiates the places much better than the raw scores. In the raw score the difference between the first pupil (raw score 48) and the second pupil (raw score 40) is only 8 (48 – 40 = 8). But in the percentile rank the difference is very wider (85 – 38 = 47) with 47. Therefore, the percentile scores/ranks distinguish pupils sharply. This procedure of estimating the norms is very much useful in schools, industries and other organizations and also in selection and placement.

The *z*-scores are used to interpret the raw scores in terms of standard deviations. In psychological and educational researches the use of *z*-scores is very rare, because of its decimal points and negative values. It is used in statistical and theoretical works only. The disadvantage involve in *z*-score is removed to some extent by transforming it to *T*-score. The same raw scores of 17 pupils are transformed to four types of norms for the purpose of comparison. The *T*-scores unlike the percentile scores/ranks do not go to the lowest number or to the highest number. For example, the first rank pupil's *T*-score is 66 and the *T*-score for the last rank is 34. Since the *z*-score and the *T*-score are based on the size of standard deviation the difference between subjects may not be too small or too high.

Stanine Scores

Test designers and testing agencies do not agree upon one norm or standard or on a common standard score unit. As result of it there are several standard score systems at present in use. Each system has its own advantages and disadvantages. The *T*-score system recommended by the Technical recommendations (1954) for psychological tests with a mean of 50 and *SD* of 10 is not without any criticisms. Now it is felt that to keep the transformed scores below 10, by which they can fit in one column in the statistical operations. The scale recommended is the stanine scale. It is a condensed form of Guilford's *c*-scale. The stanine score is very convenient for interpretation of scores. It is a type of derived score that is used in standardization of tests in recent times and also used for restandardization of old tests. Lindvall (1961) defines it as a *normalized standard score* derived from a scale having values from 1 to 9 with a mean of 5. The word stanine is derived from two words "standard" and "nine" and it is expressed as standard scale of nine units.

The stanine scale is developed on the assumption that there is normal distribution in the measurement under study. All the cases or the whole range of scores in normal distribution is divided into eleven equal intervals. This means that the range involves from minus 2.75 standard deviations to plus 2.75 standard deviations. The cases included are 99.4 per cent. The stanine scale uses only nine intervals by combining the two extreme categories at either end categories. The category one is combined with two and the category eleven is combined with ten. This makes the distribution with a total of nine score intervals. In the stanine scale the highest score is 9 and the lowest is one. The average score is five ($45 \div 9 = 5$).

The raw scores can be transformed to stanine scores by assigning percentage of cases to each stanine. In a normal distribution the percentage of cases falling in each stanine is given in Table 18.4.

Table 18.4 : Showing the percentage of cases and corresponding stanine scores

Stanines	1	2	3	4	5	6	7	8	9
Percentage of cases in each stanine	4	7	12	17	20	17	12	7	4
Cumulative Percentage		11	23	40	60	77	89	96	100

In the Normal distribution the top 4 per cent of the cases will have the highest stanine score of 9. The next 7 per cent will have the stanine score of 8. The next 12 per cent will have the stanine score of 7. The middle category with 20 per cent is assigned with a stanine score of 5 which is the average of the distribution. The other half distributions will have the stanine scores as shown in Table 18.4. The stanine scores can be interpreted with the percentage of cases. For example, a pupil has stanine score of 7. In comparing this pupil with other pupils in the group he is moderately above the mean. He has scored higher than 77 per cent of the group and has a score lower than 11 per cent and has his score as 12 per cent. One advantage of stanine score is that it can be used to compare a pupil's test score with his another test score.

Test users consider that the stanine scores have an advantage over the percentile scores/ranks. In stanine scores equal differences between scores represent equal differences in achievement, whereas in percentile scores/ranks differences in the upper and lower ends of the scale represent greater differences in achievement than the differences near the middle of the scale.

The stanine scale also has certain disadvantages. It is said that the intervals are so broad that they include large proportion of cases. The small differences in stanine scores become spurious. Pupils cannot be ranked in the correct positions. It obscures the small differences in scores which are important and have educational significances. Guilford (1987) says that the stanine scale has certain practical benefits of having nine units instead of eleven. But a rare person, one in a hundred at either end is submerged with other three adjacent three per cent. There is much discrimination between the 100th person and the next three per cent of cases. This discrimination may be as much as between any other categories. Guilford says that this loss of discrimination in the stanine scale is not tolerable.

Further, there is certain loss of information in transforming the raw scores into stanine scores. In a big group several individuals may be having different raw scores and they will be ordinarily assigned with the same stanine scores. There may be more than nine different raw scores, but they are not treated with the same differences in stanine scale. Therefore, there is a differential loss of information.

The C-Scale

The *C*-scale is also an addition to the psychological norms and scales. Guilford (1954) has developed this scale in the context of presenting normalized ranking method. The *C*-scale is a small extended form of the stanine scale. At either end of the stanine scale one interval is added to the distribution. Hence, the *C*-scale has eleven units of standard scores. Therefore, the mean of the *C*-scale is exactly 5 just like the stanine scale and its standard deviation is 2. The extreme units are 0 and 10. Hence, the class interval ranges from 0 to 10. Like the stanine scale, scores are assigned to each position. In the *C*-scale the individuals are arranged in rank order placing the highest at the top and lowest at the bottom. If the ranked group consisted of 100 persons according to the test scores the person in the highest place is assigned with a score of 10. The next three persons assigned with a score of 9. The next 7 persons are assigned with a score of 8 and the next 12 persons are assigned with 7 and so on till the last one is assigned with a zero score. The Table 18.5 gives the correctly assigned scores to the 11 units.

Table 18.5 : Showing the *C*-score and corresponding percentages of cases in distribution

	Last-Rank										I Rank
C-Scale Score	0	1	2	3	4	5	6	7	8	9	10
Percentage of cases for each unit	1	3	7	12	17	20	17	12	7	3	1
Cumulative Percentage	0	4	11	23	40	60	77	89	96	99	100

In the *C*-scale the obtained score refers to a common scale which is related to the normal distribution. If the test scores of a population yield a normal distribution then the *C*- score distribution also is normal and the units of measurement (11 units) are also equal. But this scale is not as accurate as the *T*-scale. The *C*-scale is very much useful in guidance and counselling purposes. It is also useful in rank order methods. But the use of zero score in *C*-scale causes some problems. Therefore, Guilford suggests that it is better to have the scores ranging from 1 to 11 than to have a range from zero to 10.

Profile

In the preceding sections various norms are discussed. They serve as the means of expressing the test scores in various forms for the purpose of comparison. All the norms are used for comparing one pupil with other pupils in the group on a test score. All the norms or derived scores, the percentile scores/ranks, the *z*-scores, the *T*-scores, the stanine scores and the *C*-scale may be used to compare individuals in one test score. But if individuals are to be compared on a battery of test scores we go for some other statistical treatment of the data. The one that is more convenient for this purpose is the profile.

A profile may be defined as a graphic representation of the relative magnitude of a student's scores on several test. The scores on all tests must be comparable for a profile to be meaningful. The scores should be based on the same standard scale. According to Thorndike and Hagen (1970) the set of different test scores for an individual expressed in a common unit of measure constitute his score profile. A profile is a comparison of different areas of performance made picturely clear by a graphic representation.

In the preparation of a profile, it is important that all the test scores should be converted to the same type of standard scores. If percentile score is used in the chart then the scores of all the tests should be in the form of percentile scores. Then only the scores can be meaningfully interpreted. If the profile is based on percentile scores the 50th percentile is the average of the group and this may serve as the anchorage point of the scale. In the interpretation of the scores the individual's score may be referred to this base line. This base line determines the individual's strong and weak places in the respective test performance. Therefore, the profile chart is very much exposing the individuals' level of performance in the tests.

However, care must be taken in interpreting the test scores through the profile chart. In plotting the scores in the profile the norms for all the tests must be comparable. It must be equivalent group with regard to age and grade. All the tests should be administered on a common population.

There is a tendency to see the individual's scores plotted in a profile. Particularly in guidance work, clerical work and vocational guidance and selection there is a desire to see the individual in his totality. For all these a profile approach seems to be a best method of assessing every aspect of the individual through several tests. The scores presented in a single numerical value either in the form of a summative score or in the form of a composite score or in the form of a critical score may not serve the purpose satisfactorily in comparison of individuals or in classification. Therefore, a profile is preferred as it could present a complete picture of an individual, provided suitable tests are administered and a common norm for all the scores is evolved. Textbooks in psychological statistics provide various kinds of profile charts illustrating the use of a profile. The important procedure involved in the profile chart preparation is to have a common norm for all the tests that must be comparable and plotting the scores of all the tests correctly, so as to give a full picture of the individual concerned.

19

FACTOR ANALYSIS

Introduction

The enormous amount of scientific and technological and also the social developments have enveloped the world with inumerable number of new material products for both good and bad, which resulted in making changes in the thought process and in the life-style of humans thoroughly during the course of fifty years after the World War II. People who lived before the World War II and continued to live till the present day at 1990s could very well analyse the lines on which the human life is heading at present. Change in every walk of life is a continuous process, which is taking place without any conscious attempt. Man's unquenchable thirst for new knowledge and his curiosity and anxiety to predict what would happen after 2000 A.D. in the world is a test of his intellectual potentialities and experiments. Man may state that he can predict accurately the events that are going to take place in future. He makes his claims from his past experience and incidences. But what has predicted may occur sometimes and may not occur sometimes. Man's predicting measures are always not correct. It must be acknowledged that for the most part, we have very much inadequate ideas as to what we actually must predict or measure though we have a number of methods, tools and instruments.

The astrological, astronomical and atmospheric predictions are not always correct. If they were correct we would have escaped from many miseries and unhappy occurrences. One of the reasons seem to be that we have not correctly comprehended the truth about the conditions of the atmosphere fully, the truth about the space, the truth about nuclear power, the truth about electricity and many things. But we do not reject the rights of the scientists and the astronomers' attempt to assess them. We know some of the variables of nuclear power and electricity and also we often speak in terms of them. For example, we speak of potential, resistance, inductance, velocity, nuclear fission, nuclear fusion and so on and also can state the laws of interrelationships. Similarly, in the case of human ability and human personality in general, though we could state them in certain terms, the fundamental variables or dimensions are still remained as unexplored territories in the field of psychology.

Human abilities and personality traits are multidimensional in nature. To meet the situation psychologists have been trying for a long time to formulate certain procedures and they were able to arrive at a conclusion that a statistical approach called factor analysis was the suitable method to deal with the fundamental variables and dimensions of human abilities and personality traits.

The scientists wanted to explore the smallest parts of the matter and describe the functions and the multiplicity of the phenomena, whereas, the psychologists wanted to study the smallest variables or factors that are having influence in intelligence and personality. In what behavioural aspect the functions of the variables could be observed? They tried to see this first in sensory areas. Wundt (1832-1920), (1896) had observed this problem in feeling aspect and had suggested a theory called tridimensional. Simultaneously, some of the experimental psychologists have observed this phenomenon in sensory areas like the vision, audition, touch and so on. But with regard to mental abilities and personality it is very difficult to identify the independent variable that is having influence in them. It is not possible to observe them directly and find out the trait. The possible approach is the factor analysis which was brought into the field of psychological research for assessing the unitary traits of intelligence and personality.

If we want to assess an individual's personality traits through the test scores, we must use more than one test or several tests and the test scores should be intercorrelated. By this means, it is possible to identify the tests which represent the specific variable having influence in the behaviour of the individual. But it is observed that each test may identify an underlying factor. The factor analysis simplifies the task of describing an individual by having one test to do the work of several tests. The concept, factor, is a powerful phenomenon for describing human nature.

What is a Factor?

A factor is an element or cause that contributes to a result. This is the general meaning of the concept of factor. It has several meanings. From mathematical point of view, factor refers to one of two or more integers or polynomials or sums whose product is a given integer or polynomial or sum. An integer or a polynomial or a sum can be exactly divided into another integer or polynomial. For example, 2 and 3 are the factors of 6 ($2 \times 3 = 6$). So also 1, 2, 3 are all factors of 6. Mathematically, any of the quantities that forms a product when multiplied together is a factor. It may be also stated that any of the conditions that brings about a change or result is a factor.

Psychologically a factor may be defined as the variable or trait or characteristic that causes some change in the results or contributes to a change in the results. From the factor analytic point of view a factor may be defined as an attribute (characteristic) that is drawn or inferred from the intercorrelations of a set of tests which are planned to provide a simpler and an uncomplicated explanation of what the tests possess in common.

Two Factor Theory

Early psychologists were of the view that mind is having a limited number of separate unitary powers. They were concentrating more on simple sensory and reaction tests for assessing the mental abilities. But Binet (1857-1911) strongly believed that mental ability is a complex problem and disagreed with Galton (1822-1911), Cattell (1905-1973) and others and suggested the measurement of higher functions of memory, imagination and judgment, etc. Tests of verbal, mechanical and numerical aptitudes were designed and these were found to be very much valuable in guidance and personnel classifications. Most of the mental ability test scores were found to be revealing some degree of positive correlation. It was Charles Spearman (1863-1945), (1904), an English psychologist, who was the first to observe that different tests of intelligence like the verbal, performance, numerical, etc., correlated positively with each other. On the basis of this observation Spearman concluded that all mental functions draw upon common factor, which he called the General Intelligence Factor (G). Spearman also assumed the occurrence of many specific factors, which he called *S* factors. It was also observed that because of these specific factors that the correlations between the scores of different tests do not become perfect. Spearman has observed in his studies that every ability involves some kind of *G* factor and also a collection of *S* factors. It was because of these findings, Spearman's theory is generally called 'the two factor theory'. When

Spearman has brought out his two factor theory it has aroused momentous importance in the field of mental ability testing.

Spearman's contention was that any correlation between the scores of two tests indicate a factor common to both the test scores, plus two specific factors. This may be illustrated with diagrams. Suppose, for example, we correlate the scores of two tests, *a* and *b*. We get one *G*-factor and two *S*-factors as *Sa* and *Sb*. This may be considered as tests *a* and *b* as two measures of the common factor, the *G* with two specific factors *Sa* and *Sb*. The correlation may be stated as r_{ab} since tests *a* and *b* measure the same ability (Fig. 19.1).

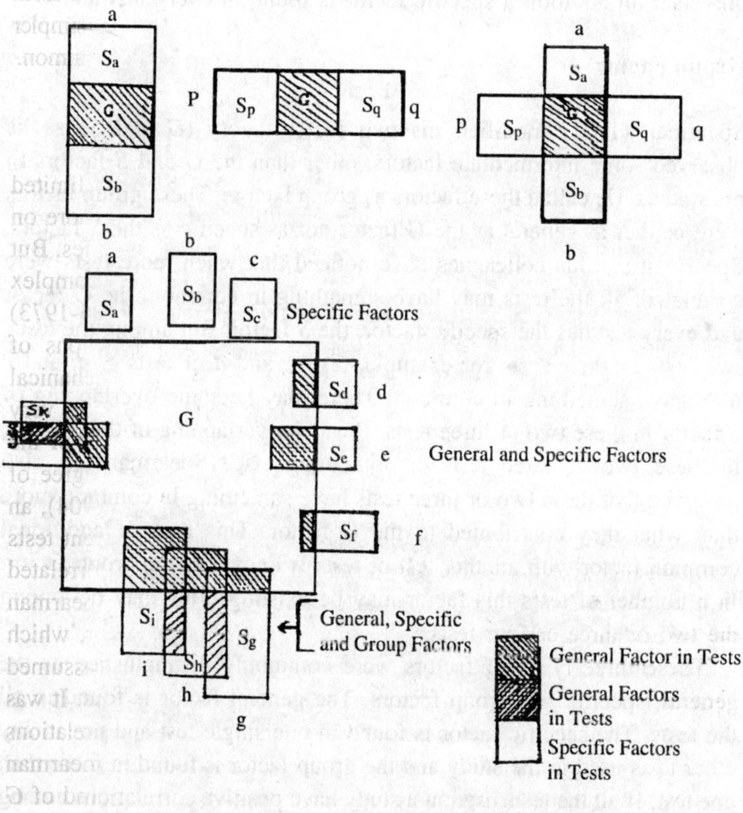

Fig. 19.1 : Possible factorial relations among the tests

If we use two more tests like *x* and *y*, the correlation yields a common factor *G* and two specific factors S_x and S_y. This may be regarded as r_{xy} as the coefficient of reliability of *a* measure *G'*. Now we have two independent measures as r_{ab} and r_{xy} and the common factors may be stated as G and *G'*. Let us now estimate the correlation between *G* and *G'*. While combining these two independent measures, we have to reduce the size or attenuate, for which we use the formula:

$$r_{gg'} = \frac{\sqrt{r_{ax}\, r_{ay}\, r_{bx}\, r_{by}}}{\sqrt{r_{ab}\, r_{xy}}}$$

Spearman draws a conclusion that the common elements are identical in all tests, as such every test is composed of the *G* factor which is a universal, in addition a specific factor is found in every test alone.

Group Factor

Spearman (1927) modified his two factor theory (*G* and *S*) as he observed some intermediate factors, other than the *G* and *S* factors in his studies. He called these factors as group factors. These group factors were neither as general as the *G* factor nor as specific as the *S* factors. Spearman and his colleagues have noticed that when more tests were correlated, all the tests may have something in common, the *G* factor and every test has the specific factor, the *S* factor. But among the tests, two tests or three tests for example, tests *c* and *d* or tests *c, d,* and *e* may have something in common. There may be some overlapping of *S* factor in these two or three tests. But is it overlapping of the *S* factor in these two or three tests or something else? Spearman has also observed that these two or three tests have something in common more than what they contributed to the *G* factor. This type of additional common factor with another test or tests is known as the group factor. In a number of tests this factor may be existing. This may be among the two or three or four tests.

These three types of factors were commonly distinguished as the general, specific and group factors. The general factor is found in all the tests. The specific factor is found in one single test and not in any other tests used in the study and the group factor is found in more than one test. If all the tests used in a study have positive correlations among them, then we can find that there is a general factor. If all the tests among them have negative correlations or zero correlations, then a general factor is not found.

Spearman and his co-workers have recognized this group factor in a number of tests like the verbal ability, numerical ability, mental speed, mechanical ability, attention and imagination. Many scholars who were working on factor theory did not agree with Spearman's group factor. Thompson (1948), who did not agree the group factor, has proposed a theory called the sampling theory. According to this theory every test goes to certain range into human abilities. Some may go widely and some may go narrowly. Therefore, the degree of correlation depends upon a part of the whole ability they have in common. Thompson accepts a general ability like the Spearman's *G*-factor, but does not accept it as a basic thing. To him it is a combination of ability elements or general ability. Similarly, for Thompson, the group factors are the collection of ability elements.

Multiple Factor Theory

There were many scholars working on factor theory both in United Kingdom and United States and all of them who were interested in factor theory seem to have accepted some form of group factor existing in their works. Those who accepted the group factor stressed that there will be always a *G*-factor in every analysis. Burt (1941), and his followers in England and Holzinger in United States have given importance to group factors and they were willing to accept the *G*-factor if it appeared in their works.

According to multiple factor theory the functions of some tests depend upon one or more common factors, and the factor loadings consist of three components namely the common factor scores plus specific component and plus an error component. According to this theory the obtained scores have some divisions which may be stated mainly as true variance and error variance. These may be also stated as components. The true variance has certain additional components. Generally they are the common factor variance and certain specific factor variance. Many tests which are having something in common contribute to the common factor variance. But the *specific factor variance* is specially the contribution of the particular test. This is considered to be *a division of the true variance*.

In every test score the variances may be present proportionately. There are concepts specifically used to indicate the proportion of the variances. They are communality, specificity and uniqueness. The common factor score variance is represented by the communality. Usually communality is symbolized by the subscript h^2. This is in the form of an equation $a_x^2 + b_x^2 + c_x^2 + d_x^2 + ... + q_x^2 = h^2$. It is the proportion, *a part of true variance*, the other part is specific variance. This may be stated either as $h_x^2 + s_x^2 = r_{tt}$ or $r_{tt} - s_x^2 = h_x^2$. The proportion of the specific variance is stated as specificity indicated by the symbol s_x^2. The other concept is the uniqueness which is symbolized by the subscript u_x^2. The variance of specificity and the error variance together is called the uniqueness. Thus, the equation is $s_x^2 + e_x^2 = u_x^2$. The uniqueness has also relationship with the communality. It is an important part of communality. Whatever a test is not having in common with other measures may be stated as uniqueness. Therefore, the equation may be stated like this: $1 - h_x^2 = u_x^2$.

The factor theory has been severely criticised by some investigators. Every test should have one factor in common. Otherwise it is not a valid test. It will not predict any criterion. But if a test measures more than one common factor which is significant then it causes some confusion. This means that the test measures something other than the purpose for which it was developed. If such incidence takes place in the measurement, how to interpret the result? Are we to accept that a test can measure more than one common factor? If that is the case the factor does not reveal the unambiguity of the test. Psychological critics like Thomas (1942) and Guilford (1954) state that the factor theory has given prominence to faulty psychological testing, because all tests measure more than one common factor. This means that each test is a unified whole, possesses a group of factors.

For example, a test measures two common factors A and B and if the factor loadings in both the common factors, A and B are equal or near equal, then it will not be possible for us to assess what amount of score has been achieved by an individual in factor A or B. This is what the criticism Thomas has levelled against the factor theory.

Thurstone (1877-1955), (1938, 1947), the great American psychologist is the pioneer of the factor analysis methods. His multiple factor analytical study is the major contribution to factor analysis. Thurstone's main aim was to describe the complex tasks as composites of simple tasks. This was done by breaking the test scores into more basic elements. Thurstone worked out to find out the group factors which are having small loadings in some performances and large loadings in other performances in his factor analysis.

For example, the Stanford-Binet memory test for sentences is depending upon verbal ability for a proportion of three by tenth (3/10) and on the memory ability for a proportion of seven by tenth (7/10). The amount of verbal ability and memory ability involved in the Stanford-Binet memory for sentences has been identified like this by factor analysis. Thurstone has identified in his studies the simple structures, which means, large number of factors were having the loadings very insignificantly or near zero values. This means that each test has just few factors. Thurstone wanted to make a mark of group factors which are having zero loadings in the tests. Then he planned to design tests which should have high loadings in one factor which he called the pure tests. According to Thurstone the tests which are of numerical operations possess this characteristics. Such tests measure the skills performed by many tests. These tests of numerical operations are almost independent of verbal reasoning and other tests of numerical abilities. Thurstone's work of the simple system of factors had great influence in the measures of abilities and classifications in the later period.

Thurstone (1938) had designed 56 tests according to the principles he formulated and administered them to the students of the University of Chicago and extracted six important factors. The factors were Verbal Comprehension (V), Word Fluency (W), Number (N), Spatial (S), Memory (M), and Reasoning (R). Thurstone called these as primary mental abilities and published the results under the title, *Primary Mental Abilities.* These six factors were extracted by the method of factor analysis. Thurstone believed that these abilities in a combined manner produce the aptitude in an individual for performing any complex mental performance. He compared the aptitude for white colour. Just like all primary colours in combination produce white colour, all the six abilities in combination produce a single mental ability called aptitude. Therefore, we may state that the general mental ability is a mixture of these six primary abilities in different proportions.

Techniques of Factor Analysis

Factor analysis is a sophisticated statistical technique designed for determining the basic elements of a measure. It has been very much useful in identifying the source of common elements that are functioning in the performance of some task in an individual or in a group. Factor analysis may be also defined as the statistical method of interpreting the scores of a set of tests by intercorrelating the test scores and bringing out the common factors that the tests are having. This method enables the researcher to find out the minimum number of factors or determinants required to explain from the intercorrelations among the test scores.

Though the method of factor analysis had its origin in the Spearman's measurement of mental abilities, it is applied to various other areas like personality measurement, assessing the role of clarity in organizational situations and so on. For example, if a researcher wants to measure the mental ability of a group of students accurately he may be using several tests devised for measuring mental abilities. One test may be overlapping another test. There may be also more overlappings among 3 or 4 tests. Therefore, what the researcher needed is to use few number of tests or some portions of tests which can measure the mental ability of the individuals or group. Factor analysis is a method which can fulfil this requirement and can discover the tests that could identify the common factors. It can also identify what are the tests that could measure other different characteristics.

In psychological studies many measuring instruments have been developed for factor analysis for determining several distinct dimensions of personality. The 16 Personality Factor Inventory developed by Raymond Bernard Cattell (1905-1973), (1950) has identified 16 separate dimensions of personality. The Minnesota Multiphasic Personality Inventory developed by Hathaway and McKinley (1943) has been factored into different forms of personality components.

A number of procedures were adapted in factor analysis. Every scholar who worked on factor analysis proposed one method of factor analysis. The basic data for factor analysis are the correlation coefficients between the test scores. Factor analysis is a procedure of evaluating the correlation values between the test scores. Since the correlation coefficients are evolved from several tests and there are a number of correlations, they are arranged in rows and columns. The arrangement is called correlation matrix.

There are five important methods of factor analysis for extracting factors. They are, the method of principle components designed by Hotelling (1933), which is supposed to be the earliest procedure of factor analysis, the second method was developed by Kelley (1935). Kelley's method is called the principle axes method, which is highly mathematical. Thirdly, the British psychologist, Sir Cyril Burt (1941) has brought out a method of factor analysis, called the summation method. Around the same time Holzinger (1941) in collaboration with Harman proposed a method of factor analysis. This method is called synthesis of factorial method, which is supposed to be the fourth method. Fifthly, a few years later the great American psychologists, Thurstone (1947) brought out his famous work on factor analysis, called multiple factor analysis. His method is called centroid method of factor analysis.

Though each one brought one method of factor analysis, two methods have something in common, but each one had its original stamp. Every author stressed one aspect as important in his method. For example, Holzinger wanted that every test should have some relationship to a *G*-factor and to a group factor. Among the methods Thurstone's method was somewhat different from other methods. He wanted that after extracting factors, the factors should be rotated. Thus, he introduced the rotation method in factor analysis. He stressed the view that a factor becomes psychologically valid and meaningful only after the rotation. But the British psychologists, Burt, did not agree with this view. Since the data were already made to undergo several steps of treatment, after extracting the factor no further treatment is required. According to him the rotated elements do not make much difference from the unrotated factors.

Choice Population for Factor Analysis

The investigators who worked on factor analysis were very curious about the population in which they were conducting the investigation for the treatment of factor analysis. Mostly they preferred a homogeneous population from the factor analysis point of view. Certain controls were imposed for obtaining a clear structure of factors. The population variable may interfere as common factors in the individual differences in which the researcher or the investigator is interested. Therefore, it is felt that certain variables should be controlled. These variables are supposed to be common in many things. The variables such as age, gender, educational levels, income levels and type of home town. There may be one or two variables highly correlate with these common variables and create an illusion as if they were actually correlating with these variables though they were not really correlating. This type of function of the variables may affect the results very much. Therefore, it is better to control certain variables for factor analysis treatment of the data.

It is also observed that investigators in order to have the sample larger, may combine some other samples drawn from other areas in factor analysis. But it is better to test whether the samples added differ from the first sample very much. There is a statistical procedure to test this. The means of the samples may be worked out separately and compared with the means of the first sample, and the sample means which are having much difference may be dropped. But some authors like Guilford (1954) suggest that some adjustment can be made and samples may be combined. But this type of adjustment in the samples may not be very much reliable.

Techniques of Factor Analysis

Factor analysis is made according to some specific techniques. The experimental data can be dealt with three different ways in factor analysis. But what technique the experimenter applies to his data should be planned in advance. There are three techniques of factor analysis. They are the *P* technique, *Q* technique and the *R* technique. In the *P* technique mostly the occasions are used as the population. The occasions may be the number of times, or the number of days, or weeks or months. It may also be ten times or ten days, or ten weeks, or it may be 20 or 30 times or days. Whatever, the variable is, the number of occurrences is taken as the sample. It may be behavioural occasions. Intercorrelations are worked out between the occasions of a person. The correlation table is stated as the occasion matrix which is often used in psychophysical methods. Therefore, it is concentrating on the scores of the individual person, obtained from him on different times.

This is similar to *R* technique in which the intercorrelations are made among the large number of tests or items. The analysis is also similar to *R* technique. The *P* technique factor analysis was first used by Cattell (1946) in the study of personality measurement. Later on Cattell and Rhymer (1947) applied the *P* technique in the determination of the individual traits. Another investigator, Watson (1951) applied this technique in assessing factors in single person's behaviour.

The *Q* technique of factor analysis starts with a table of matrix of intercorrelations among the scores of the persons. The rows and columns consisted of the intercorrelations of scores between persons. The number of rows and the number of columns are the number of persons to whom the tests were administered. One specific feature of the *Q* technique is that the number of variables are in large number, larger than the number of persons. The number of persons may be 20 or 30 or 40. But the number of variables used in the experiment may be 100 or even 200. Instead of working out the intercorrelations among 100 or 200 test scores in the *Q* technique we are correlating the scores between the number of persons, whether it is 20 or 30. This technique is mostly used in the field of clinical psychology for identifying the dominant syndromes. Enormous amount of data or information may be collected about the persons to whom the tests were administered.

The *R* technique of factor analysis is very widely used by the investigators. In the *R* technique of factor analysis, intercorrelations are worked out between the scores of the tests. The correlation matrix consisted of the same number of rows and columns of the tests used in the investigation. The correlations are worked out between columns or in other words between tests in *R* technique. The *R* technique of factor analysis is more practical and most of the investigators apply only the *R* technique in their studies. In the *R* technique of factor analysis it is possible to identify the factor that is common to the whole sample and also the factors that are specific. This technique can be applied in various behavioural aspect, like the study of intelligence, personality, interest and aptitudes, attitudes and prejudices and also in almost all the areas.

The Correlation Matrix

The method of factor analysis begins with the correlation matrix. The correlation matrix has many rows and columns as the number of tests used in the study. The components of correlation matrix are the intercorrelations among the tests. The size of the correlation matrix depends upon the number of tests an investigator uses for his study or the number of items of a scale or an inventory. In a correlation matrix the correlation coefficients may range from –1.00 to +1.00, if maximum positive and negative correlations occurred. The steps involved in factor analysis can be illustrated with a model problem. An investigator who is interested in applying factor analysis to his research work wanted to extract two or three factors by using six tests. He administered these six tests to a small group of 20 university students who were studying in the postgraduate classes, and collected the data. This is the first step. The next step is to work out the intercorrelations among the test scores. Since there are six tests, the intercorrelations are $\frac{n(n-1)}{2} = \frac{6 \times 5}{2} = 15$. The correlations could be worked out with the help of a calculating machine or a computer.

Now there are 15 correlation coefficients obtained from 6 tests. These correlations are arranged in a matrix. While arranging these 15 correlations in a matrix, we must bear in mind that they should find places on the upper right portion of the matrix (Table 19.1). The matrix is only half filled, the lower left portion of the matrix is empty. The lower left portion should be filled with the same data, or the correlations. Now only 30 cells are filled up and the diagonal cells are not filled (Table 19.1). But filling or not filling of the diagonal cells depends upon the procedure we use for factor extraction. The matrix consists of 36 (6 × 6) cells. The 15 cells of the upper right portion and the 15 cells of the lower left portion are filled. But the diagonal cells should be filled with an element called communality of the test. The communalities are the values that are in agreement with the correlation coefficients in the cells. Several procedures are suggested to fill up the diagonal cells. One of the procedures is to put the highest possible correlation of 1.00 in the diagonal cells and proceed on according to the procedure for extracting factors. This is applied by Fruchter (1954). Thurstone, (1947) suggested to use gussed communalities in the diagonal cells. Though this procedure may cause some errors, it can be used without damaging the factors extracted. But one of the best procedures is to insert the highest correlation coefficient (*r*) found in every column as suggested by Guilford (1954). That is, each diagonal cell can be filled up with the correlation coefficient that is highest in that specific column and the procedure may be continued. Having thus the correlation matrix is completed Thurstone's (1947) complete centroid method may be employed. But Thurstone uses the value of the highest correlation of the column not in the diagonal cell. He uses it as the *D* value and adds it with the total '*s*' value.

Factor Extraction

The basic elements for factor extraction are the correlation coefficients in a matrix form. There is also another method of factor extraction by using the covariances and some investigators use this method. But the use of correlation matrix for factor analysis is easy and it is also safe method in extracting factors. Therefore, in the model illustration the correlation matrix method is used.

Generally if one wants to apply factor analysis in his study, he should use more number of tests, whether it is to measure intelligence, or personality, or aptitude or attitudes. Thurstone (1938) used 56 tests and identified seven distinct factors. Cattell (1950) used 16 tests in his studies, which he called 16 Personality Factor Test. Wechsler (1939) devised the Adult Intelligence Scale with 11 tests for factor analysis, six of them were verbal comprehension and five of them performance type. The M.M.P.I. developed by Hathaway and McKinley (1942) consisted of nine sub-tests. Thus, tests were designed for applying factor analysis to identify the factors.

A mental ability test consisted of six sub-tests was used by an investigator, referred above, for a research work. The data of this study were taken for illustrating a problem. The tests were, (1) Reading Ability, (2) Verbal Comprehension, (3) Arithmetic Reasoning, (4) Mechanical Ability, (5) Spatial Relations, and (6) Assembly. These tests were administered to a small group of 20 university students and the raw scores were obtained by scoring the tests according to the procedure.

I Stage : Extraction of the First Centroid Factor Loadings

Since the correlations are the basic data in factor analysis, the first step is to work out the intercorrelations among these six tests. The intercorrelations yielded 15 correlations and they are presented in Table 19.1. It is observed that there are substantial correlations among the six tests of a battery. The correlation between the first-two tests are relatively high. They appear to define a factor of ability to deal with verbal in relationship. The tests 3 and 4 are also having substantial correlation and both are numerically related. Between tests 5 and 6 the correlation is very substantial and both tests involve in something of visual perception.

The aim of factor analysis is to identify a small number of factors that are underlying in the performance of the individuals in common. The factors account for the relationship among the test variables. Each test is having a loading in each of the factors and we are trying to identify the pattern of factor loading that is psychologically meaningful. Though the correlation coefficients themselves show the type of relationship between two tests, we do not know what amount of common relationship each test is having with every other test in the battery. The factor analysis alone would reveal the amount of relation each test has with other tests.

The assumption, by administering these six tests, is that the first-two tests are dealing with language ability, the second-two tests are dealing with mathematical ability and the last two tests are referring to spatial relations and assembly. Persons who have better verbal comprehension also supposed to possess better reading ability and they may be falling behind in other tests. Similarly persons who are high in arithmetic reasoning also possess high mechanical ability. Persons who score high marks in mathematics in schools prefer to take engineering courses of mechanical type. Therefore, those, who score high in arithmetic reasoning also possess mechanical ability high. Thirdly, those who score high in spatial relation tests may easily identify the relationship between parts and space and they also supposed to possess the ability to assemble parts which are spread out in scattered form. Therefore, persons who score high in spatial relations may also score high in assembly tests. The assumptions are tested in this study by applying the factor analysis method.

Table 19.1 : Correlation matrix showing the intercorrelations of 6 Mental Ability Tests

S. No.	*Tests*	*Reading Ability*	*Verbal Compre.*	*Arithme. Reasoning*	*Mech. Ability*	*Spatial Relation*	*Assembly*
1.	Reading Ability	—	0.60	0.10	0.08	0.13	0.12
2.	Verbal Comprehension	—	—	0.12	0.14	0.12	0.11
3.	Arithmetic Reasoning	—	—	—	0.45	0.05	0.10
4.	Mechanical Ability	—	—	—	—	0.10	0.11
5.	Spatial Relations	—	—	—	—	—	0.43
6.	Assembly	—	—	—	—	—	—

The second step in this procedure of factor analysis is to have a full correlation matrix. The correlations presented in Table 19.1 are occupying only the upper right portion of the matrix. To have full correlation matrix the lower left portion also should be completed. Therefore, the matrix is completed by entering the same correlations, found on the upper right portion of the matrix, in the lower left portion correspondingly as shown in Table 19.2.

Since the number of correlations are 15 for six tests and when the same are entered on the lower left portion of the matrix in Table 19.2, there are 30 entries of correlations. But in a 6 × 6 correlation matrix there are 36 cells. The six diagonal cells are empty. In the procedure of factor extraction these diagonal cells should be filled up with some values. The ways of filling up the diagonal cells are already stated in the preceding section. Up to this stage the procedure of factor analysis is common to all the factor analysis methods. Procedures vary when values are inserted in the diagonal cells.

We shall take up first the complete centroid method of factor extraction from the correlation matrix in Table 19.3, step by step. The problem now is the estimation of communality for each test. The simplest method of estimating the communality is to use the highest correlation coefficient in each column of the matrix. According to the complete centroid method the communality is not inserted in the diagonal cells. Later on it is added to the sum of the elements of the column.

Table 19.2 : Correlation matrix showing the intercorrelations of six mental ability tests with lower left portion of the matrix completed

S. No.	*Tests*	*Reading Ability*	*Verbal Compre.*	*Arithme. Reasoning*	*Mech. Ability*	*Spatial Relation*	*Assembly*
1.	Reading Ability	—	0.60	0.10	0.08	0.13	0.12
2.	Verbal Comprehension	0.60	—	0.12	0.14	0.12	0.11
3.	Arithmetic Reasoning	0.10	0.12	—	0.45	0.05	0.10
4.	Mechanical Ability	0.08	0.14	0.45	—	0.10	0.11
5.	Spatial Relations	0.13	0.12	0.05	0.10	—	0.43
6.	Assembly	0.12	0.11	0.10	0.11	0.43	—

The third step is to sum up the columns and entering them in the last row against S_1. The rows are summed up and entered in the last column under check sum. The sums of rows and columns should tally and check marks are to be made on each of the sum. Further the sum of the sums should agree. In this problem every sum tally and the sums agree and it is 5.52.

In the fourth step the *D* values (communalities) are entered just below S_1 values in the row against *D*. The '*D*' value is the diagonal value. The *D* value is the highest correlation in each column. In this problem, the D value in the first column is 0.60, in the second column it is 0.60, in the third 0.45, in the fourth 0.45, in the fifth 0.43, and in sixth column it is 0.43. These values are summed up and entered in the last column just below the sum of S_1 and it is 2.93.

Then in the fifth step, the S_1 values and the D values in the rows are added and entered in the row just below the D value against $S_1 + D = E$. The total of these two values is E. In this problem the E values are 1.63, 1.69, 1.27, 1.33, 1.26, and 1.30.

The sixth step is to add the sums of S_1 and the D values and it is 8.48 in this problem. The sums of these two values are checked and they tally. The sums of S_1 and D values constitute the E values. The E values of all the columns are summed up and it is stated as the T value. The T is the grand total. In this problem it is 8.48.

In the seventh step the root of T is worked out. Since the formula to work out the factor loadings is: $a_1 = \frac{Ej}{\sqrt{T}} = mEj$, the root of T is worked out and it is, $\sqrt{T} = \sqrt{8.48} = 2.912044$. In the equation a_1 is referring to the factor loadings, E is referring to the total value of $S_1 + D$ values, j refers to the tests (in this problem there are 6 tests). T is the grand total, and m is the value of 1 divided by the root of $T\,(\frac{1}{\sqrt{T}})$.

In the eighth step the products of T are checked. The obtained root of T is 2.912044 and the $\frac{1}{\sqrt{T}}$ is 0.343401 in this problem. It is important to check whether the root of T is equivalent to the equation mT. In this problem the value of mT is 2.912044 which is equivalent to the root of T, 2.912044. Check marks are made on these three values, T, $\frac{1}{\sqrt{T}}$ and mT and these values should be given in five decimal digits.

In the ninth and last step the complete centroid factor loadings are extracted. From the above illustration we could understand that it is done by multiplying every E value by the root of T ($\sqrt{T}$) and dividing the product by T (8.48). But this work is simplified by using a constant m through the equation $\frac{1}{\sqrt{T}} = m$. In this problem the constant m is 0.343401. The formula for extracting the factor loadings at this state is: $a_1 = mE$. Therefore, we are multiplying each E by the constant m. Thus on multiplying the two variables m and E, we have the factor

Table 19.3: Showing the extraction of the first centroid factor loadings from the correlation matrix

Tests	1	2	3	4	5	6	*Check sums*	
1	—	0.60	0.10	0.88	0.13	0.12	1.03 √	
2	0.60	—	0.12	0.14	0.12	0.11	1.09 √	
3	0.10	0.12	—	0.45	0.05	0.10	0.82 √	
4	0.08	0.14	0.45	—	0.10	0.11	0.88 √	
5	0.13	0.12	0.05	0.10	—	0.43	0.83 √	
6	0.12	0.11	0.10	0.11	0.43	—	0.87 √	
S_1	1.03	1.09	0.82	0.88	0.83	0.87	5.52 √	
D	0.60	0.60	0.45	0.45	0.43	0.43	2.96 √	
$S_1 + D = E$	1.63	1.69	1.27	1.33	1.26	1.33	8.48 √	$= T \cdot \sqrt{T} = 2.912044$
$mE = a_1$	0.560	0.580	0.436	0.457	0.433	0.446	2.912 √	$= 1/\sqrt{T} = 0.343041 = m$
								$= mT = 2.912044$

loadings for the test 1, (m E), 0.343401 × 1.63 = 0.560. In the same manner the factor loadings are worked out for all the six tests and we have first centroid factor loadings as, 0.560, 0.580, 0.436, 0.457, 0.433, and 0.446.

These factor loadings are presented in the last row of the Table 19.3. It is important to check the factor loadings by an equation of the root of T. The sum of the factor loadings should agree with $\sqrt{T}$. In this problem the root of T is 2.912044 with five decimal places and the sum of the factor loadings is 2.912 in three decimal places. Since these two figures agree the factor loadings are correct.

II Stage : Extraction of the Second Centroid Factor Loadings

In the second stage of this factor analysis, we are preparing for the extraction of the second factor loadings. The first step is to prepare a table for working out the residuals of the first factor loadings. Table 19.4 is containing fully the residuals of the first factor. In Table 19.4, at the left end, in the first column the first factor loadings a_1 are entered in the order from 1 to 6 of the tests and also the tests in column 2. The sum of the first factor loadings of the six tests is given with positive signs at the bottom of the column and indicated by V_1. Then at the top of the *Table 19.4*, in the row, K_1 is entered. Then the first factor loadings and the sum of them are entered with minus signs. The sum of the first factor loadings is indicated with $-V_1$. Now we are working out the residual matrix of the first factor in Table 19.4. The Table 19.4 now is with entries

Table 19.4 : Showing the procedure of computation of the residuals of the first factor and forming the matrix of the residuals for the extraction of the second factor

$-k_1$		–.560	–.580	–.436	–.457	–.433	–.446	$-V = -2.912$
a_1	Test	1	2	3	4	5	6	
.560	1	—	(–.325 + .60) .275	(–.244 + .10) –.144	(–.256 + .08) –.176	(–.242 + .13) –.112	(–.250 + .12) –.129	
.580	2	(–.325 + .60) .275	—	(–.253 + .12) –.133	(–.265 + .14) –.125	(–.251 + .12) –.131	(–.259 + .11) –.149	
.436	3	(–.244 + .10) –.144	(–.253 + .12) –.133	—	(–.199 + .45) –.250	(–.189 + .05) –.138	(–.194 + .10) –.094	
.457	4	(–.256 + .08) –.176	(–.265 + .14) –.125	(–.199 + .45) .250	—	(–.198 + .10) –.098	(–.204 + .11) –.094	
.433	5	(–,243 + .13) –.112	(–.251 + .12) –.131	(–.189 + .05) –.139	(–.198 + .10) –.098	—	(–.193 + .43) .237	
.446	6	(–.250 + .12) –.129	(–.259 + .11) –.149	(–.194 + .10) –.094	(–.204 + .11) –.094	(–.193 + .43) .237	—	
$2.912 = V_1$								

of the first factor loadings in column as a_1 and in row as K_1 with minus signs as shown in the Table 19.4. The sum of the factor loadings in the row is with minus values with an indication of $-V_1$.

The second step is to multiply a_1 and $-k_1$ $[(a_1)(-k_1)]$ and to add with the element in every respective cell of the correlation matrix of Table 19.3. But it should be remembered that the variable a_1 is positive and K_1 is negative. The procedure is actually deducting the products from the elements of the correlation matrix. But we are not doing like that, since the products are minus values. Therefore, in order to avoid any confusion in dealing with the figures, it is suggested to add the products with every element found in Table 19.3. But the residual matrix is not going to be fully with minus values, even though the factor loadings (K_1) in the row are having the minus signs. If the figure in the correlation matrix (Table 19.3) is higher than the product of $(a_1)(-K_1)$ then the residual value is plus, without minus sign. Therefore, the residual matrix is with both positive and negative values. Let us work out one example for both positive and negative in the residual matrix. In Table 19.4, the factor loadings for the test 1, a_1 and $-k_1$ are multiplied, $-0.580 \times 0.560 = -0.325$. The value in the correlation matrix for the test 1 is 0.60. When we add the product of $(a_1) \times (-k_1)$ -0.325 to the correlation 0.60 we get the residual value as $(-0.325 + 0.60) = +0.275$. We get the residual value positively because the correlation value is greater than the product of $(a_1)(-k_1)$.

But if the correlation in Table 19.3 is smaller than the product of $(a_1)(-k_1)$ as in the case of test 3 ($r = 0.10$), the product is $-0.436 \times 0.560 = -0.244$. By adding this value -0.244 with 0.10, $(-0.244 + 0.10) = -0.144$, which is the residual value. The residual values presented in Table 19.4 are the residuals of the first factor loadings. These are worked out very carefully and the illustration is made step by step to avoid any confusion in evolving the residual matrix for extracting the second factor. Table 19.4 shows how the residuals of the first factor loadings are computed and obtained. The procedure involves the multiplication of a_1 and k_1 and adding the respective correlation in Table 19.3, $(a_1)(-k_1) + r_{11}$.

Thus, every element of the correlation matrix (Table 19.3) finds a place in different form in Table 19.4 and in Table 19.5. In this problem it can be observed that there are 6 elements with positive signs and rest of them with negative signs.

The third step is to prepare a table for presenting the residuals of the first factor for extracting the second factor loadings. The residuals are presented in Table 19.5. In Thurstone's complete centroid factor analysis the procedure of extracting the second factor is little different from the procedure of the first factor. The residuals are entered in Table 19.5 in rows and columns, leaving the diagonal cells empty. Then the rows and columns are summed up and the sums are checked and check marks are made. The sums of the rows and columns have negative signs. The sums of the columns are entered in the row against S_2 and the sums of the rows are entered in the column under check sums.

The fourth step involves entering the S_1 values from Table 19.3 in the row against S_1 just below S_2 values in Table 19.5.

In the fifth step the K_1 and V_1 products are worked out ($K_1 \times V_1$) and entered in the row against K_1 V_1. The K_1 values are the first factor loadings given in the top row with minus signs in Table 19.5. The V_1 value is the sum of the first factor loadings of the six tests given at the bottom of the column a_1 in Table 19.5. Let us work out one example of this. The K_1 value for the first test is –0.560 and the V_1 value is 2.912. When these two are multiplied ($-0.560 \times 2.912 =$ -1.630), the value we get is –1.630 and this is entered in the row against K_1 V_1 just under S_1 values. The value of K_1 varies test to test, but the V_1 is not varying, it is a constant. This is worked out for all six tests and the values are summed up and entered under check sum and it is –8.479 and the same value is checked by the equation $(-V)$ (V), $-2.912 \times 2.912 = -8.479$. Since both the values are agreeing the calculations are correct.

The sixth step is the working out of the values of K_1^2 of every test. This is obtained by squaring the first factor loadings presented in the first row at the top of the *Table 19.5*. One *example* is: $-0.560 \times -0.560 = 0.314$, and this is entered in the row against K_1^2 for all the six tests, following the same procedure the row K_1^2 is completed and the sum is checked and verified by the equation Σa_1^2. This is by squaring the value under a_1 and summing up. This value (Σa_1^2) is 1.435, the same value of K_1^2 and a check mark is made on it.

The seventh step is summing up the values of S_1, K_1V_1 and K_1^2. The value of these three elements are added and entered in the row against

Table 19.5 : Showing the first factor residuals and the extraction of the second factor

k_1		–.560	–.580	–.436	–.457	–.433	–.446	$-2.912 = -V_1$	
a_1	Test	1	2	3	4	5	6	Check sum	
.560	1	—	.275	– .144	– .176	– .112	– .129	– .286 √	
.580	2	.275	—	– .133	– .125	– .131	– .149	– .263 √	
.436	3	–.144	– .133	—	.250	– .139	– .094	– .260 √	
.457	4	–.176	– .125	.250	—	– .098	– .094	– .243 √	
.433	5	–.112	– .131	– .139	– .098	—	.237	– .243 √	
.446	6	–.129	– .149	– .094	– .094	.237	—	– .229 √	
$V_1 = 2.912$	S_2	–.286	– .263	– .260	– .243	– .243	– .229	–1.524 √	
	S_1	1.030	1.090	.820	.880	.830	.870	5.520 √	
	$K_1 V_1$	–1.630	–1.689	–1.270	–1.331	–1.261	1.298	–8.479 √	$(-V_1)(V_1) = -8.479$ √
	K_1^2	.314	.336	.190	.209	.187	.199	1.435 √	$\Sigma a_1^2 = 1.435$ √
$S_1 + K_1 V_1 + K_1^2$		–.286	– .263	– .260	– .242	– .244	– .229	–1.524 √	
S^2		–.286	– .263	– .260	– .243	– .243	– .229	–1.524 √	
$-S_2/2 = A$		.1430	.1315	.130	.1215	.1215	.1145	.762 √	
+1		(.143)	.407	– .014	– .055	.0095	– .0145	.476 √	
$+2 = B$		(.418)	(.407)	– .147	– .180	– .1215	– .1635	.213 √	
$B(-2) = C$		– .836	– .814	.294	.360	.2430	.3270	– .426 √	$\Sigma \mid C \mid = 2.874$ √
D		– .275	– .275	.250	,250	.237	.237		$\Sigma \mid D \mid = 1.524$ √
$C + D = E$		–1.111	–1.089	.544	.610	.480	.564	4.398 √	$= T$ 4.398 √
$mE = a_2$		–.530	– .519	.259	.291	.229	.269	2.097	$\sqrt{T} = 2.09714$ $1/\sqrt{T} = 0.47683 = m$ $mT = 2.097141$ √

$S_1 + K_1 \ V_1 + K_1^2$ just under K_1^2 values. The sum of this value is also checked in this step that $\Sigma K_1 V_1$ equals $(-V)$ (V) and both have the same value.

In the eighth step it is to be verified whether the sum of S_2 (−1.524) is equal to the sum of $S_1 + K_1 V_1 + K_1^2$. These two values are equal and it is −1.524 and entered in Table 19.5 in the respective cells in the last column and they are checked and marked.

In the ninth step we are determining the value of A and it is by using the formula $\frac{-S_2}{2}$. The values of $-S_2$ are divided by two and entered in the row against A and the values are turned to be positive.

In the tenth step the values in the row against A are to be changed to either to positive or negative without changing the numerical values. This is done by finding out the test which is having highest positive value and entering the value in the cell immediately below that value and making it within parentheses. In this problem the highest positive value is for test 1. This test number is written as +1 just below A in the row and the value 0.143, is written within parentheses against +1. This means that the sum of residual values of test 1 is not changed numerically, but only the sign is changed as minus (Table 19.5).

In the 11th step the residual values of the test 1 are added with every other test values which are written against A in the row. In the case of test 1, since the same value (0.143) is accepted we are not adding this value with test 1. But we are adding with other test values. Thus in the case of test 2, the A value is 0.1315, with which the value 0.275 is added and we get 0.407 and the same is written against +1. For example, in the case of test 3, the A value is 0.130 and the residual value is −0.144. By adding −0.144 to 0.130, we get −0.014. The same procedure is carried out for all other tests and the values are entered in the respective cells. At this stage all the positive signs are not changed. The highest positive 0.407 is for test 2 and therefore, test 2 is reflected by putting this value 0.407 within parentheses and retaining the numerical value. It is reflected in the case of test 1 as it is having positive sign. Therefore, the same value (0.407) is written immediately below within parentheses. Then the residual value of 0.275 is added to 0.143 and we get the value 0.418 within parentheses. Similarly with test 5, by adding the residual value of the test 1, −0.112 with 0.0095, we get the value as −0.1215. Now we can see after two reflections all the six tests turned minus values. The *values within the parentheses are not minus.* Therefore, we stop the reflection and the last obtained value is B, it is +2 = B. These reflected values are summed up and entered in the last column under check sum (Table 19.5).

In the 12th step we are obtaining the C value by multiplying the B values by –2. Since both the values (the B values and –2) are having minus signs, the products become positive values. But the values within the parentheses are considered to be positive. For example, in the case of test 1, the B value is 0.418. This is multiplied by –2 and we get the value –0.836, similarly in the case of test 2, it is –0.814 (0.404 × (–2) = –0.814). But in the case of other four tests the products are positive. Then the C-values are summed up with signs for checking up and they are also summed up disregarding the signs and entered at the end of the row. Now we find that the sum of C-value is the double of the B-value in every test.

Then in the 13th step the D-values are entered just below the C- value for each test. The D-values are the diagonal values (communalities) the highest correlation in each column. Then the C- and the D-values are added in rows and also summed up and entered at the last column. The sum of these two values is E and entered in the row against E. The sum of C and D values are checked with the sum of E values and they tally. The sum of the E value is the T, the grand total of C and D.

In the 14th step we are extracting the second centroid factor loadings, for which we must obtain the value 'm'. The m is a constant obtained from the grand total T. The formula is : $1/\sqrt{T} = m$. In this problem the T is 4.398 and root of which is $\sqrt{4.398} = 2.09714$. $1/\sqrt{T} = 1/2.09714 = 0.47683 = m$. The mT is equal to $\sqrt{T}$. Therefore, it is checked and found to be correct in this problem it is: 0.47683 × 4.398 = 2.09714.

In the 15th step the second factor loadings are extracted by applying the formula $mE = a_2$. The m which we have obtained in the preceding step is used to multiply each E for obtaining the factor loadings a_2. The m is 0.47683. For example, the E of test 1, is –1.111 is multiplied by (m) 0.47683 and we get the loading –0.530 for test 1. The same procedure is followed for all other tests and we have the second centroid factor loadings for the six tests as, –0.530, –0.519, 0.259, 0.299, 0.229, and 0.269. Lastly the sum of a_2 is checked, whether the sum of a_2 equalizes $\sqrt{T}$. The sum is 2.097 and it tallies with $\sqrt{T}$ (Table 19.5).

III Stage : Extraction of Third Centroid Factor Loadings

Now we are arriving at third stage of factor analysis for extracting third centroid factor. The first step is to evolve the second factor residuals and present them Table 19.6. The values required for this is the residuals of the second factor in a matrix. A workable table is prepared and the factor loadings in rows and columns are entered. The top row is K_2 and the column one is a_2. In the second column the tests are written. In row K_2 the second factor loadings are written with signs reversed. In column one under a_2 the factor loadings are written as obtained without changing the signs (Table 19.6).

In the second step the second factor residuals are worked out and entered in the respective cells leaving the diagonal cells empty. The steps in computing the second factor residuals are the same as those we have described in the preceding section for the first factor residuals. We can show one *example* here. For example, in Table 19.6, the diagonal cells are left blank. We go to the next cell, a_{21} and K_{22} are multiplied $(-0.530 \times 0.519) = -0.275$ and this value is entered immediately below the test 2. The corresponding residual value of the second factor (Table 19.5) is 0.275. This is added to the product –0.275. After adding (–0.275 + 0.275) the value is 0.000. This is written in the row against test 1, in column 2. The same procedure is followed in completing the second factor residual matrix. Thus in Table 19.6 the residuals of the second factor are presented.

The third step is to prepare a workable Table 19.7 and entering in it all the residual values computed. Accordingly they were worked out and presented in Table 19.7, in the respective cells. The same steps that we described in the extraction of the second factor are followed in the extraction of the third factor loadings at all levels. One important point is that the signs in rows and columns should be as stated in tables.

The last row of the Table 19.7 brings out the third centroid factor loadings. The factor loadings are, 0.042, –0.042, –0.423, –0.414, 0.431 and 0.404. In the last column the checkings and tallyings are made as shown in Table 19.5.

Stage IV : Extraction of the Fourth Centroid Factor Loadings

In the third centroid factor, the first two loadings are very negligible, but in other four tests the loadings are very substantial. Further the

Table 19.6 : Showing the procedure of computing the residuals of the second factor loadings

K_2		.530	.519	–.259	–.291	–.229	–.269 $-V_2$ = .001
a_2	Test	1	2	3	4	5	6
–.530	1	—	(–.275 + .275) .000	[.137 + (–.144)] –.007	[.154 + (–.176)] –.022	[.121 + (–.112)] .009	[.142 + (–.129)] .014
–.519	2	(.275 + .275) –.000	—	[.134 + (–.133)] .001	[.151 + (–.125)] .026	[.119 + (–.131)] –.012	[.139 + (–.149)] .014
.259	3	[.137 + (–.144)] –.007	{.134 + (–.133)] .001	—	(–.075 + .250) .175	[–.059 + (–.139)] –.198	[–.069 + (–.094)] –.164
.291	4	[.154 + (–.176)] –.022	[.151 + (–.125)] .026	[–.075 + (.250)] .175	—	[–.067 + (–.098)] –.165	[–.078 + (–.094)] –.172
.229	5	[.121 + (–.112)] .009	[.119 + (–.131)] –.012	[–.059 + (–.139)] –.198	[–,067 + (–.098)] –.165	—	(–.061 + .237) .175
.269	6	[.142 + (–.129)] .014	[.139 + (–.149)] –.009	[–.069 + (–.094)] –.164	[–.078 + (–.014)] –.172	(–.062 + .237) .175	—

V_2 = .001

Table 19.7 : Showing the residuals of the second factor and the extraction of the third factor

K_2		–.530	–.519	–.259	–.291	–.229	–.269	$-.001 = -V_2$	
a_2	Test	1	2	3	4	5	6	Check sum	
–.530	1	—	.000	–.007	–.022	.009	.014	–.006 √	
–.519	2	.000	—	.001	.026	–.012	–.009	.006 √	
.259	3	–.007	.001	—	.175	–.198	–.164	–.193 √	
.291	4	–.022	.026	.175	—	–.165	–.172	–.158 √	
.229	5	.009	–.012	–.198	–.165	—	.175	–.191 √	
.269	6	–.014	–.009	–.164	–.172	.175	—	–.156 √	
$V_2 = .001$	S_3	–.006	.006	–.193	–.158	–.191	–.156	–.698 √	
	S_2	–.286	–.263	–.260	–.243	–.243	–.229	–1.524 √	
	$K_2\ V_2$	.000	–.000	–.000	–.000	–.000	–.000	.000	$(-V)\ (V) = .000001$ √
	K_2^2	.281	.269	.067	.085	.052	.072	.826	$\Sigma\ a_2^2 = .826$ √
	$S_2 + V_2 + K_2^2$	–.005	.006	–.193	–.158	–.191	–.157	–.698 √	
	S_3	–.006	.006	–.193	–.158	–.191	–.156	–.698 √	
	$-S_3 - S_2 = A$	.003	–.003	.0965	–.079	.0955	.078	.349 √	
	+3	–.004	–.002	(.0970)	.254	–.1025	–.086	.1565 √	
	+4	–.026	.024	(.2720)	(.254)	–.2675	–.258	–.0015 √	
	$B + 2$	–.026	(.024)	(.2730)	(.276)	–.2795	–.267	.0005 √	
	$-2B = C$	.052	–.048	–.546	–.552	.5590	.534	–.001	$\Sigma \mid C \mid = 2.291$ √
	D	.022	–.026	–.196	–.175	.1980	.175		$\Sigma \mid D \mid = .794$ √
	$C + D = E$	.074	–.074	–.744	–.727	.757	.709	3.085	$= T$ 3.085 √
	$mE = a_3$	.042	–.042	–.423	–.414	.431	.404	1.756	$\sqrt{T} = 1.75641$ $1 \div \sqrt{T} = 0.569341 = m$ $mT = 1.75641$ √

Table 19.8 : Showing the procedure of computing the residuals of the third factor

a_3	K_3	–.042	.042	.423	.414	–.431	–.404 $V_3 = .002$
	Test	1	2	3	4	5	6
.042	1	—	.002 + .000 .002	.018 + (–.007) .011	.017 + (–.022) –.005	–.016 + .009 –.007	–.017 + .014 –.003
–.042	2	.002 + .000 .002	—	–.016 + .001 –.017	–.017 + .026 .009	.018 + (–.012) .006	.017 + (–.009) .008
–.423	3	.018 + (–.007) .011	–.018 + .001 –.017	—	–.175 + .175 .000	.182+ (–.198) –.016	171 + (–.164) .007
–.414	4	.017 + (–.022) –.005	–.017 + .026 .009	–.175 + .175 .000	—	.178 + (–.165) .013	–.167 + (–.172) –.005
.431	5	–.018 + .009 –.009	.018 + (–.012) .006	.182 + (–.198) –.016	.178 + (–.165) .013	—	–.174 + .175 .001
.404	6	–.017 + .014 –.003	.017 + (–.009) .008	.171+ (–.164) .007	.167 + (–.172) –.005	–.174 + .175 .001	—
$-V_3 = -.002$							

residuals seem to be still strong. Therefore, we are going for the fourth factor stage. The first step is to prepare a workable Table 19.8 and to compute the residuals of the third factor. In Table 19.8, the first row in filled with K_3 values. These are the third factor loadings and they are written with reverse signs. In column under a_3 the factor loadings are entered as obtained with the same signs. The sums of the rows and the columns are checked at the end of the row and at the end of the column.

Then in the next step the same procedure is followed in computing the residuals of the third factor, leaving the diagonal cells empty. For example here, K_{32} and a_{31} are, 0.042 and 0.042 and when these two are multiplied the product is 0.002, the second factor residual is 0.000 and, therefore, the value to be entered in the cell is $(0.042 \times 0.042 = 0.002)$, and when the product and the residual are added it is $0.002 + 0.000 = 0.002$. Another *example*, $K_{33} \times a_{32} = 0.423 \times (-0.042) = 0.016$. The residual value from Table 19.7 is 0.001 and when these two added the product is $-0.016 + 0.001 = -0.017$. Thus, the third factor residuals are computed for all six tests and presented in Table 19.8.

In the third step, a workable Table 19.9 is prepared and the products are entered and the third residual matrix is evolved in Table 19.9. In the first row of Table 19.9, K_3 the third factor loadings are entered with change of signs and the sum also is entered as $-V_3$. In the column below a_3 the factors as extracted with the same signs are entered and the sum as V_3. All the 30 cells are filled up with third factor residuals, leaving the diagonal cells empty. The sums of the rows and columns are checked up. The sum of the sums also checked and tallied (Table 19.9). It is observed that the sums in rows and columns have become very weak and vanishing and the sum of the sums 0.004 is almost in vanishing state. But let us proceed further to see what is happening with this problem. The same procedures described above are followed for extracting the factor loadings. In the last row the fourth factor loadings are entered. The $mE = a_4$, the fourth loadings are, –0.072, 0.108, –0.106, 0.068, 0.139 and 0.016. The factor loadings are very low and negligible. The highest loading is for the test 5, 0.139. They are all near zero and, therefore, we stop at this stage.

Now let us test the proportion of variances contributed by the centroid factors.

The centroid factor analysis has yielded four factors. The first factor (Table 19.10) has positive loadings in all items. The loadings range from

Table 19.9 : Showing the residuals of the third factor and the extraction of fourth factor

K_3		–.042	.042	.423	.414	–.431	–.404	.002 = V_3	
a_3	Test	1	2	3	4	5	6	Check sum	
.042	1	—	.002	.011	–.005	–.009	–.003	–.004 √	
–.042	2	.002	—	–.017	.009	.006	.008	.008 √	
–.423	3	.011	–.017	—	.000	–.016	.007	–.015 √	
–.414	4	–.005	–.009	.000	—	.013	–.005	.012 √	
.431	5	–.009	.006	–.016	.013	—	.001	–.005 √	
.404	6	–.003	.008	.007	–.005	.001	—	.008 √	
–.002 = V_3	S_4	–.004	.008	–.015	.012	–.005	.008	.004 √	
	S_3	–.006	.006	–.193	–.158	–.191	–.156	–.698 √	
	$K_3 V_3$	.000	.000	–.001	–.001	.001	.001	.000	$(-V_3)(K) = .000004$ √
	K_3^2	.002	.002	.179	.171	.186	.163	.699	$\Sigma a_3^2 = .699$ √
	$S_3 + K_3V_3 + K_3^2$	–.004	.008	–.015	.012	.004	.008	.013 √	
	S_4	–.004	.008	–.015	.012	.005	.008	.014 √	
	$-S_4 + 2 = A$	.002	.004	.0075	–.006	–.0025	–.004	–.007 √	
	+3	.013	–.025	(.0075)	–.006	–.0185	.003	–.022 √	
	$B + 1$	(.013)	–.019	(.0185)	–.011	–.0275	.000	–.026 √	
	$-2B = C$	–.026	.038	–.037	.022	.055	.000	.052	$\Sigma\|C\| = .178$ √
	D	–.011	.017	–.017	.013	.016	.008	.026	$\Sigma\|D\| = .082$ √
	$C + D = E$	–.037	.055	–.054	.035	.071	.008	.260	$= T$ .260 √
	$mE = a_4$	–.072	.108	–.106	.068	.139	.016	.509	$\sqrt{T} = 0.50990$ $1/\sqrt{T} = 1.96117 = m$ $mT = 0.50990$ √

0.433 to 0.580. There are heavy loadings in the first two items which are dealing with reading ability and verbal comprehension. The loadings on other four items are moderate. But all of them are above 0.400. Since all the six items are having the loadings above 0.433, a common factor runs through all the items or tests. This means all the individuals in the group have the ability in all the areas to some extent. The students who scored high in the first-two tests also scored moderately high in other four tests. Therefore, there is a common factor operating through all the six tests which we may call it as general intelligence, the *G* factor according to Spearman. The first factor accounts for 47.42 per cent of the total variance, whereas, the variance attributed to the factors second, third and fourth are 27.63, 23.23 and 1.72 percentages respectively. The first factor has exhausted significant common variance and the factors two and three are bipolar having both positive and negative loadings and exhausted about 50 per cent of the total variance.

Table 19.10 : Showing the factor matrix of the factor loadings for the Group (*N*-20)

Test		A_1	A_2	A_3	A_4	A_1^2	A_2^2	A_3^2	A_4^2	h^2
Reading Ability	1	.560	–.530	.042	–.072	.314	.281	.002	.005	.602
Verbal Comprehension	2	.580	–.519	–.042	.108	.336	.269	.002	.012	.619
Arithmetic Reasoning	3	.436	.259	–.423	–.106	.190	.067	.179	.011	.447
Mechanical Ability	4	.457	.291	–.414	.068	.209	.085	.171	.005	.470
Spatial Relation	5	.433	.229	.431	.139	.187	.052	.186	.019	.454
Assembly	6	.446	.269	.404	.016	,199	.072	.163	.000	.434
				ΣA_k^2		1.435	.826	.703	.052	3.016
						47.42%	27.63%	23.23%	1.72%	100%

It is observed that four of the tests 3, 4, 5 and 6 have almost the same amount of loadings ranging from 0.229 to 0.269, in factor II (A_2) which include arithmetic reasoning, mechanical ability, spatial relations and assembly. But, since the factor loadings are not substantial, all below 0.269, it is not possible for us to ascertain what exactly these loadings reveal at this stage. Probably, we may know more about these after rotation. But we cannot rule out the existence of something in common among these four tests. The loadings in factor III (A_3) are zero for the first-two tests (0.042 and –0.042) and also negative loadings for tests 3 and 4. But the tests 5 and 6 have some substantial loadings (5, 0.431, 6, 0.404). This factor seems to reveal the perceptual ability of the students, since the tests are dealing with spatial relations and assembling of parts. In assembling various parts in an instrument or in a machine, perception of spatial relations is a vital factor. Therefore, these two abilities seem to be going together in this experiment. At the same time the loadings in arithmetic reasoning and mechanical ability there are negative and zero loadings indicating that these items react in the reverse. This means that there is no relationship between the first-two and the second-two tests and between the second-two and the third-two tests, in factor III (A_3). The third factor has exhausted 23.23 per cent of the total variance.

Factor IV (A_4) is a residual factor, having no significant loading in any of the tests. All the tests have zero loadings except the test 5. Since this factor has been extracted from the vanishing third factor residuals the loadings are almost zero, insignificant and also negative. Strictly speaking we should have stopped with factor III (A_3). The fourth is extracted only for illustration, to show what is happening in factor extraction, if extraction is made from the vanishing residuals. Therefore, according to the principles of factor analysis we should stop after extracting the III (A_3) factor. We are not taking the fourth factor into consideration and we are dropping it at this stage without rotation.

Methods of Rotation

We shall now rotate the first-three factors. Rotation is another statistical treatment applied in factor analysis. This is intended to provide psychological meaning to the factor loadings. Actually there were difference of opinions with regard to rotation of the extracted factors. Exponents in the field of factor analysis expressed different views. Thurstone (1940), Guilford (1954) Reyburn and Taylor (1943) in the United States, in contrast to Burt (1941) in United Kingdom believed that it is necessary to rotate the factors for obtaining the psychological meaning since the meaning of the factors is not clear in the unrotated factors. They are of the view that it is a way of objective approach. Therefore, the extracted factors are rotated.

There are two important methods of rotation. What method is to be applied depends upon the nature of the data. The methods are: (1) the graphic orthogonal rotation, and (2) oblique rotation. The orthogonal rotation was suggested by Zimmerman (1946) and it is named after him as Zimmerman's graphic rotation method. In this method the axes are maintained at 90° in rotation. It should be observed that while the rotation is made, maximum number of plots of the tests are brought within the axes.

The oblique rotation is little different from the orthogonal rotation. In the oblique method the axes need not be always at 90° and each axis may be rotated independently. The oblique is made in 3 or 4 ways. Among them only the radial method is most generally applied in research works.

In the referred problem the graphic orthogonal method of rotation is employed for rotating three factors. This method is more suitable for this type of study.

Rotation of the Factors: Orthogonal Rotation

The first step in applying orthogonal rotation is to have a workable graphic sheet with marked reference axes as $-A_0O + A_0$ and $+B_0O - B_0$ for the first rotation. The first rotation for our problem is in the plane of A_0OB_0 and the centroid axes are A_0 and B_0. Since three factors with substantial loadings are extracted, three rotations are made in this problem. Hence three graph sheets are needed for three rotations. The first rotation is made by plotting the projections of the factors A_0 and B_0. The second rotation is made by plotting the projections of A_1 and C_1 and the third rotation is made plotting the projections of the factors B_2 and C_2. The axes are always maintained at 90° in all rotations.

The problem which we are illustrating here has three factors with substantial loadings. The fourth factor is not having any substantial loading and it has mostly zero loadings. Therefore, we are omitting the fourth factor from the process of rotation. If there are only two factors with substantial loadings, then we can rotate the factors only once. It is called two dimensional rotation. Since there are three factors extracted with substantial loadings in our problem, we must have different technique for rotating the three factors. The technique we are using for this is called three dimensional rotation. But since both techniques are important and every one should know them we are employing both the techniques for illustration.

Two Dimensional Rotation

The two dimensional rotation needs only two factors extracted from the tests. Out of the three factors we are going to rotate, we shall take up the first-two factors for the two dimensional rotation, factors *A* and *B*.

The first step has been already stated in the preceding section to have a workable graphic sheet and to make the preparations needed. In the frame of reference of the centroid axes (Fig. 19.2) mention the centroid axes A_0 and B_0 in the plane. Now the graph sheet is ready.

In the second step the projections of the factors *A* and *B* are plotted. While plotting one should see that the test vectors or the projections of the two factors are meeting at a point either in A_o axis or in B_o axis.

For example, the loading of the first test in the first factor is 0.560 and for the same first test in the second factor the loading is –0.530. These two projections are plotted in the frame of reference and the loading 0.560 should go into the area of +*A* axis and the vector or the loading –0.530 should go into the area of –*B* axis. The +*A* axis and the –*B* axis meeting in an area where the two loading are plotted and the point of place is labelled as 1 and a circle is made. The second test loadings for the factor *A* is 0.580 and for the factor *B* is –0.519. These two vectors are plotted at a place where a label 2 is made and a circle is made. But in the case of test 3, both the vectors or loadings are positive. Therefore, these two loadings plotted at a point of area which is positive for both *A* and *B* frames of reference and the point is labelled as 3, and a circle is made. In the same manner all the six test vectors of the factors A_0 and B_0 are plotted in the frame of reference.

The third step is to rotate the axes A_0 and B_0 in such a way that most of the vectors of the tests plotted in the frame of reference must come into the right angle between the new axes $+A_1$ and $+B_1$. In this process we must see that there are as many zero loadings as possible or brought near the axes. In the process of rotation for maintaining the 90° always, it is better to use the set squares. One set square is fixed exactly on the line –*AO*+*A* horizontally and another set square on the vertical line, so that a right angle is formed. Then both the set squares are slowly moved or rotated on the axes A_0 and B_0 towards right in the same 90° position in such a way that more number of plottings are brought within the right angle and some of the plottings with zero value if possible. The rotation is stopped when as many plottings as possible brought within the right angle. Then two lines are drawn joining *O* with the frames of reference. Now we find two new axes and they are indicated with the letters A_1 and B_1. As far as possible it is better to have the two new axes, A_1 to be nearer to A_0 and B_1 nearer to B_0 (Fig. 19.2).

But sometimes the new axis B_1 may go nearer to the axis A_0 and A_1 may go nearer to the axis B_0. In such case it is to be labelled differently. The B_1 is labelled as $+A_1$ and A_1 is labelled as $+B_1$. Actually if B_1 goes below the frame of reference of A_0 it has to be labelled as $-B_1$. That is what normally done in the preparation of Direction Numbers (L_{01}) and in Direction Cosines (T_{01}) or in the Transformation Matrix (T_{o1}). But in the graph it is not labelled like that. For example, in the case of the problem we are discussing and illustrating here, in

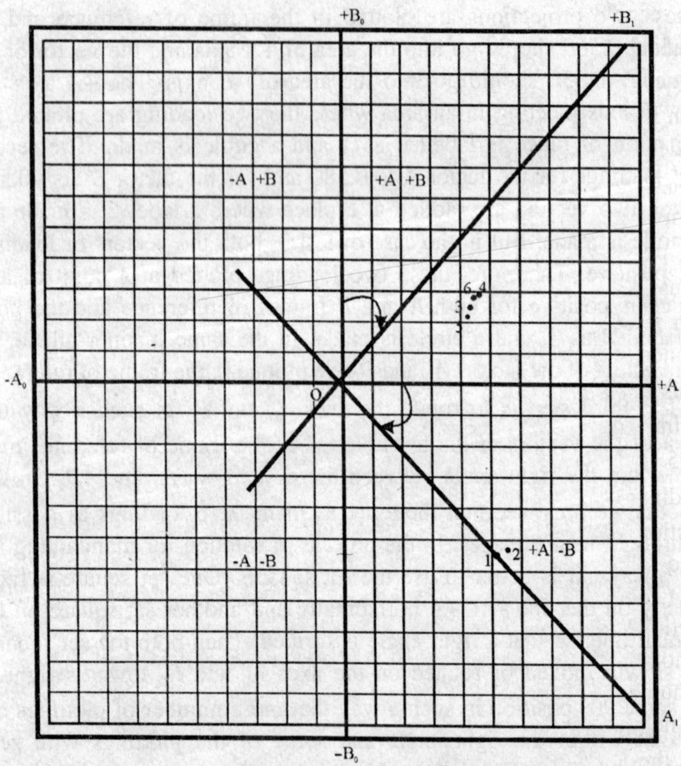

Fig. 19.2: Showing the plottings of the six test vectors in the plane *AOB* of the Centroid Factor Axes *Ao* and *Bo* and the first Orthogonal Rotation to the new positions of the Axes A_1 and B_1 of the Factors I and II.

all the three rotations the axis B_1 (Fig. 19.2), C_2 (Fig. 19.3) and C_3 (Fig. 19.4) fall within the frames of reference of $-B$ and $-C$ and we treat them in the same manner with negative value. But we labelled them as $+B_1$, $+C_2$ and $+C_3$.

The fourth step is working out the Direction Numbers for the two new axes, A_1 and B_1. Each axis has two direction numbers. Thus, there are four direction numbers such as A_0, B_0, A_1 and B_1. The direction numbers for this problem are worked out for the two dimensional rotation and the rotated factor matrix is evolved and presented here (Table 19.11). The determination of direction number and the direction of cosines is stated as the transformation matrix (T_{o1}). This matrix is used for multiplying the centroid factor matrix for obtaining the rotated factor matrix. The direction numbers are obtained from rotation of the test vectors in the planes of the centroid axes A_1 and B_1. The direction numbers are in two pairs used for coordinating the centroid factors and the new positions of the vectors.

In the fifth step the direction numbers of the transformation matrix are determined. After plotting the test vectors and rotation the positions of A_0 and A_1, the positions of B_0 and B_1 are observed. Axis A_o is moved downward to a new position A_1 and axis B_0 moved on the right to a new position B_1 (Fig. 19.2). In the case of our problem, let us first discuss axis A_1. The distance of axis A_1 in the direction of A_o is 1.00 from left to right. At the same time the corresponding coordinating distance in the direction of B_0 is –0.95, as the axis B_o is moved on the minus frame of reference.

Similarly in the case of B_1 the distance in the direction of $+B$ is +0.95 between B_o and B_1 in the plane and the corresponding coordinating distance in the direction of B_o from A_0 is 1.00. These direction numbers are presented in the form of matrix and the same is called transformation matrix and the elements in this matrix are stated as l (Table 19.11). In the orthogonal two dimensional rotation the direction numbers are only two. But the same numbers are presented as another pair of numbers in the reverse order. In each pair there is one number with positive sign and the same number in another pair is with negative sign (Table 19.11)

In the sixth step the direction cosines of the transformation matrix (T_{o1}) are computed from the direction numbers. In this process it is important to check the columns and rows of the transformation matrix

Table 19.11 : Two dimensional orthogonal rotation of the axes in the problem illustrated

(a) Centroid Factor Matrix (C)

	A_0	B_0	h^2
1.	0.560	–0.530	0.594
2.	0.580	–0.519	0.605
3.	0.436	0.259	0.257
4.	0.457	0.291	0.293
5.	0.433	0,229	0.239
6.	0.446	0.269	0.271

(b) Direction of Numbers (L_{01})

	A_1	B_1
A_0	1.00	0.95
B_0	–0.95	1.00
Σl^2	1.9025	1.9025
$\sqrt{\Sigma l^2}$	1.3793	1.3793
$1/\sqrt{\Sigma l^2} = D_{01}$	0.725	0.6888
	–0.6888	0.725

(c) Transformation Matrix (T_{01})

	A_1	B_1	Σl^2
A_o	0.7250	0.6888	1.0000
B_o	0.6888	0.7250	1.0000
Σl^2	1.0000	1.0000	

(d) First Rotated Factor Matrix (O_1)

	A_1	B_1	h^2
1.	0.771	0.001	0.594
2.	0.778	0.023	0.605
3.	0.138	0.488	0.257
4.	0.131	0.526	0.293
5.	0.156	0.464	0.239
6.	0.138	0.502	0.271

as the sum of columns and rows should be identical. In this problem the same is worked out and found correct (Table 19.11). The elements of the direction numbers are squired and summed up (Σl^2). In this problem it is 1.9025. Then the root of this value is obtained $\sqrt{\Sigma l^2}$ and it is 1.3793. Then this is made to divide the value of 1, ($1\sqrt{\Sigma l^2}$). The divided value is labelled as D_{01} and the value is 0.725. This D_{01} (0.725) is the coordinating value for 1, if so, what is the coordinating value for direction number 0.95? The formula for determining this value is $1(0.725) \div 0.95 = 0.6888$. These values are posted in the respective axes (Table 19.1) and the T_{01} matrix is formed and the rows and columns are checked. The four elements in the transformation matrix are used as the coordinating values between the centroid factor matrix and the rotated factor matrix.

The seventh step is to compute the rotated factor loadings and to form the matrix. This is obtained by multiplying the loadings of the centroid factors by the elements of the transformation matrix, the direction cosines (T_{01}). In this procedure two loadings of the test 1, of the centroid factor matrix, one under A_0 and one under B_0 are multiplied by all four elements of the transformation matrix and the products are entered in the rotated factor matrix (O_1), one product under A_1 and the other product under B_1. The A_0 product goes to the corresponding axis A_1 and B_0 product goes to the respective axis B_1. Two examples are worked out here.

The centroid factor loading of test 1, under factor A_0 is 0.560 and the loading of the same test under factor B_0 is –0.530. The respective multiplying direction cosines are 0.7250 and –0.6888. Therefore, the computation is (0.560) (0.7250) + (–0.530) (–0.6888) = 0.771 and this value is entered under A_1 against test 1. In the case of the first loading of the second factor the direction cosines are 0.6888 and 0.7250. The computation is : (0.560)(0.6888) + (–0.530) (0.7250) = 0.001. This is written under B_1 against test 1. The same procedure is followed in computing the centroid factor loadings by the coordinating numbers for forming the rotated factor matrix.

The eighth step is working out the communalities of the tests. This is worked out separately for the centroid factor matrix and for the rotated factor matrix. It is expected that the communalities of both matrixes should agree and tally. This is one way of checking whether the computed communalities in the rotated factor matrix agree with the communalities of the centroid factor matrix. The figures in the last column of the matrixes are almost the same (Table 19.11).

Interpretation of Unrotated Factors (Two Dimensional)

The centroid factor analysis has yielded two factors. The first factor has positive loadings in all the six tests. The loadings range from 0.433 to 0.580. These loadings are substantial and it may be stated that a common factor runs through all the tests as revealed by the scores of a group of 20 students. But however, the first-two tests differentiate from the other four tests, indicating that the group is more proficient in abilities concerning language. The first factor (*A*, 1.435) accounts for 63.5 per cent of the total variance (2.253).

The second factor is a bipolar factor, having two negative loadings in tests 1 and 2. Though the loadings are negative they are high and significant. In the case of other four tests the loadings are not substantial, they are within 0.269. A projection or factor loading, according to Thurstone (1940) less than 0.40 is not significant and the naming of the factor cannot be made. This factor accounts for 36.5 per cent of the total variance. Unless rotation is made it is not possible to state any thing more than this.

Interpretation of Rotated Factors (Two Dimensional)

The first factor has four projections almost at the same level, 0.138, 0.131, 0.156, and 0.138 and they are nearly in vanishing state. The lowest projection (0.131) is for the mechanical ability test and the next lowest is for arithmetic reasoning (0.138) and for assembly (0.138). In spatial relation test the loading is 0.156. But for the tests 1 and 2 the projections are (0.771 and 0.778) highly significant. These tests 1 and 2 are dealing with reading ability and verbal comprehension, mainly with language or verbalization. It appears that those who are proficient in verbalization or language fall below in other abilities. The other abilities in this study are dealing with mainly numerical and spatial problems. This factor (tests 1 and 2) accounts for 56.5 per cent of the total variance. The loadings in the first-two tests after rotation synchronize with the loadings of unrotated centroid factor. At both levels the loadings are substantial. Therefore, the projections confirm result that these two tests are mainly concerned with verbalization or language.

The factor after rotation has vanishing loadings in the first-two tests. The projections are almost nil (1, 0.001 and 2, 0.023). But in the case of other tests, 3, 4, 5 and 6 the projections range from 0.414 to 0.526 which are substantial and significant. This factor for four projections accounts for 43.5 per cent of the total variance. Actually the results of this problem seem to be more reliable after rotation since in the first factor the tests 1 and 2, have highly significant loadings and the other tests, 3, 4, 5, and 6 are almost in vanishing state. But in the second factor tests 1 and 2 are almost vanished and tests 3, 4, 5, and 6 have substantial loadings. This much of clear results are obtained in rotating the centroid factor. Thus, the rotation of the centroid factors in this problem provided better results, in other words, better psychological meaning to the problem than the unrotated factors.

Three Dimensional Rotation of the Centroid Factors

With regard to rotation there is not much difference between the two dimensional rotation and three dimensional rotation. In two dimensional rotation there are only two factors and the rotation is made once. The axis A_0 and the axis B_0 are made to move on the plane to new positions A_1 and B_1. But in three dimensional rotation we will have to make three rotations, A_0 and B_0 are made to move to the new positions of A_1 and B_1, secondly, A_1 and C_1 to move to the new positions of A_2 and C_2 and thirdly B_2 and C_2 to move to the new positions of B_3 and C_3.

Secondly, the elements in transformation matrix, the direction numbers (L_{01}) and the direction cosines (T_{01}) are only four in two dimensional rotation. But in three dimensional rotation the direction numbers (L_{01}) are four and the direction cosines (T_{01}) are nine. Thus, the elements of direction numbers are four like the two dimensional but, the elements of direction cosines are nine, three for each axis, A_0, B_0, C_0. Though only two factor loadings are plotted in the frames of reference for rotation, any one of the three is kept as constant. Thus, the three dimensional rotation varies from the two dimensional when we prepare the transformation matrix (T_{01})) after determining the direction numbers (L_{01}), particularly when the direction cosines are determined.

First Rotation: Three Dimensional

The three dimensional rotation requires a minimum of three factors extracted from the correlation matrix. In Table 19.12, the centroid factor matrix with three factors are presented, A_0, B_0 and C_0. On plotting the projections of the factors in A_0 and B_0 axes in the Fig. 19.3, the axes are rotated and moved to the new positions A_1 and B_1. This task is performed in the same manner as we did in the two dimensional rotation. Since we can use only two factors for a rotation, we are plotting the projections of two factors at a time. The same procedure is used in determining the direction numbers (D_{01}). After determining the direction numbers (Table 19.12) in the same manner as we did in the previous section (two dimensional) we go for determining the direction cosines (T_{01}). The direction number (D_{01}) is the same (0.725) as for two dimensional. Here we are using the same data for three dimensional rotation also. The two direction cosines are also the same (0.7250 and 0.6888), but we add the value of 1.00 for the factor C, since we did not rotate factor C and it is constant. The rows and columns of the direction cosines are summed up and checked the elements of the direction cosines (T_{01}).

Table 19.12 : Three dimensional rotation of the Axes in the problem illustrated

(a) Centroid Factor Matrix (C)

	A_0	B_0	C_0	h^2
1.	0.560	–0.530	0.042	0.594
2.	0.580	–0.519	–0.042	0.606
3.	0.436	0.259	–0.423	0.436
4.	0.457	0.291	–0.414	0.464
5.	0.433	0.229	0.431	0.424
6.	0.446	0.269	0.404	0.434

(b) Direction Numbers (L_{01})

	A_1	B_1
A_o	1.00	0.95
B_o	–0.95	1.00
Σl^2	1.9025	1.9025
$\sqrt{\Sigma l^2}$	1.37931	1.37931
$1/\sqrt{\Sigma l_2} = D_{01}$	0.725	0.725

(c) Transformation Matrix (T_{01})
(Direction Cosines)

	A_1	B_1	C_1	Σl^2
A_o	0.7250	0.6888	0.0000	1.0000
B_o	–0.6888	0.7250	0.0000	1.0000
C_0	0.0000	0.0000	1.0000	1.0000
Σl^2	1.000	1.000	1.0000	

(d) First Rotated Factor Matrix (O_1)

	A_1	B_1	C_1	h^2
1.	0.771	0.001	0.042	0.595
2.	0.778	0.023	–0.042	0.606
3.	0.138	0.488	–0.423	0.436
4.	0.131	0.526	–0.414	0.464
5.	0.156	0.464	0.431	0.424
6.	0.138	0.502	0.404	0.434

The first-three dimensional rotation is made for factors A_0 and B_0. Therefore, the next step is to form the first rotation factor matrix (O_1), by multiplying the centroid factor loadings by the coordinating elements in the direction cosines matrix. Since we did not rotate the factor C_0, the elements in A_0 and B_0 are used in the computation. The computation procedure also is the same as for two dimensional. The first two projections of test 1, one under A_0 and one under B_0 are multiplied by the coordinating elements 0.725, –0.6888, 0.6888, and 0.725. Let us work out one example: (0.560) (0.725) + (–0.530) (–0.6888) = 0.771, which is written under A_1 against test 1. Similarly, the same loadings are multiplied by the corresponding coordinating elements (0.560) (0.6888) + (–0.530) (0.725) = 0.001. This is written under B_1 against test 1.

Another example for the projections of test 2 is : (580) (0.725) + (–0.519) (–0.6888) = 0.778. This is noted under A_1 against test 2, for the direction cosines B_1, (0.580) (0.6888) + (–0.519)(0.725) = 0.023. This is entered under B_1 against test 2. Thus, the first rotated factor matrix is completed in the same manner and we have the Matrix O_1 with A_1 B_1 and C_1 factors. The values under A_1 and B_1 are computed through the coordinating elements. But the factor C is not rotated. According to the procedure the loadings of factor C are repeated in the matrix as they are without any change in the values and all the three factors are labelled as A_1, B_1 and C_1.

The next step is to compute the communalities to test whether the rotated factor matrix agree with the communalities of the centroid factor matrix. With regard to the problem, the communalities of both the matrixes agree with each other (Table 19.12). In orthogonal rotation this check is necessary to find out whether the computations are correct.

Second Rotation : Three Dimensional

In making second rotation of the centroid factors the same procedure is followed step by step. As stated earlier the first step is to prepare the graph with plane and axes for the factor to be rotated (Fig. 19.3). In the second rotation the projections of the factors A_1 and C_1 are plotted in the graph. In factor C_1 three of the projections are having negative values. Whatever they are, they all plotted according to the procedure in Fig. 19.3. The orthogonal rotation is made by using two set squares at the centre of the graph in a right angle triangle and then the set squares are moved clockwise on the right and stopped the rotation where more number of the plottings are brought within the right angle triangle and also on making plottings to reach the zero point. Now the axes A_1 and C_1 reached new positions A_2 and C_2 and the axes are marked as A_2 and C_2. A straight line is drawn from the central point O to the axis A_2 and from O to the Axis C_2.

In the next step the direction numbers and the distance between the axes A_1 and A_2 and the distance between C_1 and C_2 are determined for working out the transformation matrix.

In the second rotation the direction number (L_{12}) Table 19.14(b) are 1.00 and –0.35 and 0.35 and 1.00. Using these direction numbers the D_{12} is worked out. The procedure to find out the D_{12} is the same as explained in the preceding section. The elements of the direction numbers are labelled as l. In this problem the D_{12} is 0.944. This is for A_2. The distance in the direction of A_2 is 1.00, the corresponding value in the direction of C_2 is –0.35. For the axis C_1 the direction numbers are 0.35 and 1.00. The D_{12} value for the distance of 1.00 is 0.944. Then what is the D_{12} value for –0.35? It is –0.33. With these elements the transformation matrix or the direction cosines is formed for A_2, B_2 and C_2. The rows and the columns

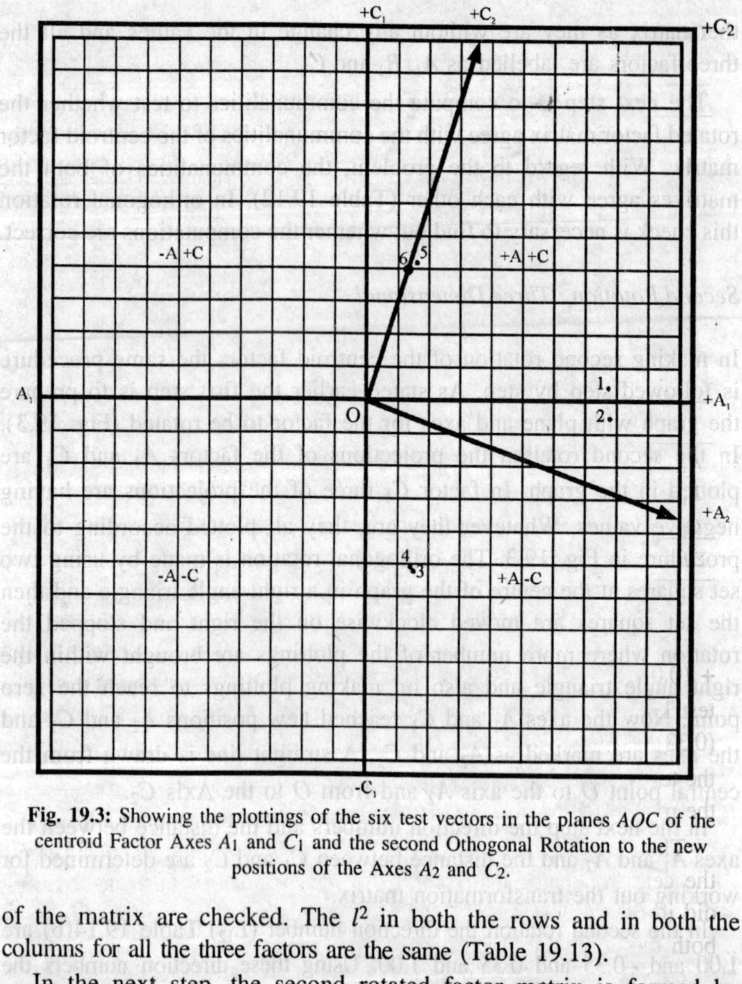

Fig. 19.3: Showing the plottings of the six test vectors in the planes *AOC* of the centroid Factor Axes A_1 and C_1 and the second Othogonal Rotation to the new positions of the Axes A_2 and C_2.

of the matrix are checked. The l^2 in both the rows and in both the columns for all the three factors are the same (Table 19.13).

In the next step, the second rotated factor matrix is formed by multiplying the centroid factor matrix and transformation matrix. Here factor B_1 is repeated as it is not rotated without changing the values. The projections of the factors A_1 and C_1 are multiplied by the corresponding coordinating elements. Let us see how the operation provides the elements to the rotated factor matrix O_2 multiplying the centroid matrix by the coordinates. The projection under A_1 is 0.771 and under C_1 is 0.042 for the test 1. Each one is multiplied by the corresponding elements of T_{12}, and added and entered under A_2 and C_2. Thus we have (0.771) (0.944)

Table 19.13 : Three dimensional rotation of the axes in the problem illustrated

(a) Centroid Factor Matrix (C)

	A_1	B_1	C_1	h^2
1.	0.771	0.001	0.042	0.595
2.	0.778	0.023	–0.042	0.606
3.	0.138	0.488	–0.423	0.436
4.	0.131	0.526	–0.414	0.464
5.	0.156	0.464	0.431	0.424
6.	0.138	0.502	0.404	0.434

(b) Direction Number (L_{12})

	A_2	C_2
A_1	1.00	0.35
B_1	–0.35	1.00
Σl^2	1.1225	1.1225
$\sqrt{\Sigma l^2}$	1.05948	1.05948
$\sqrt{\Sigma l^2} = D_{12}$	0.94385	0.94385

(c) Direction Cosines (T_{12})

	A_2	B_2	C_2	Σl^2
A_1	0.944	0.000	0.330	1.000
B_1	0.000	0.000	0.000	1.000
C_1	–0.330	0.000	0.944	1.000
Σl^2	1.000	1.000	1.000	

(d) Second Rotated Matrix (O_2)

	A_2	B_2	C_2	h^2
1.	0.714	0.001	0.293	0.595
2.	0.748	0.023	0.218	0.606
3.	0.269	0.488	–0.354	0.435
4.	0.259	0.526	–0.348	0.464
5.	0.005	0.464	0.458	0.424
6.	–0.003	0.502	0.426	0.434

+ (0.042) (–0.33) = 0.714. This value is entered under A_2 against the test 1. Similarly with the other pair of direction cosines, we have : (0.771) (0.33) + (0.042) (0.944) = 0.293. This value is entered under C_2 against the test 1. The same procedure is adapted throughout the computation and the rotated factor matrix O_2 is evolved (Table 19.13).

The next step is to work out the communalities and check whether the centroid factor matrix and the rotated factor matrix agree, as we did in the case of the first rotation in this problem. It is observed that both the communalities (h^2) are almost the same.

Third Rotation: Three Dimensional

The third rotation is made from the same data of the problem, in continuation of the second rotation described above. In the previous two rotations factors A_0 and B_0, A_1 and C_1 were rotated. In the third rotation we are using the factors B_2 and C_2. In the first rotation factor C_0 was not rotated and it was repeated. In the second rotation factor B_1 was not rotated and repeated. But in the third rotation factor A_2 is not rotated and it is kept constant by repeating it.

Therefore in the third rotation the centroid factor loadings of B_2 and C_2 are plotted in the graph prepared for this (Fig. 19.4). There are two zero loadings in factor B_2 and two negative loadings in factor C_2. Since there are two zero projections and two negative projections the plottings take abnormal places and the figure may be little different from the other two figures. It can be observed in the graph that the axes B_3 and C_3 cut the plane at the very close point near B_2 and C_2. The plottings of the projections are made in the same manner as we did in the previous cases. The axes B_2 and C_2 have taken only very small turning in rotation, the distance between B_2 and B_3 and also the distance between C_2 and C_3 is very small. Therefore, based upon the new positions of the axes B_3 and C_3 the transformation matrix is worked out using the direction numbers (L_{23}). Following the same procedure described in the first rotation the D_{23} is computed. The D_{23} for the problem at this stage is 0.9994 and based upon this the direction cosines (T_{23}) computed and the elements are presented in the matrix C of the Table 19.14. Thus, we have the coordinating elements 0.9994 and –0.035 and also the corresponding elements 0.35 and 0.9994 (Table 19.14).

The next step is multiplying the centroid factor matrix by the corresponding element from the direction cosines and evolving the third rotated factor matrix. The same procedure described above is followed step by step. Let us compute two examples for the test 1, using the

Table 19.14 : Three dimensional rotation of the axes in the problem illustrated

(a) Centroid Factor Matrix (C)

	A_2	B_2	C_2	h^2
1.	0.714	0.001	0.293	0.595
2.	0.748	0.023	0.218	0.607
3.	0.269	0.488	–0.354	0.435
4.	0.259	0.526	–0.348	0.465
5.	0.005	0.464	0.458	0.424
6.	–0.003	0.502	0.426	0.433

(b) Direction Numbers (L_{23})

	B_3	C_3
B_2	1.000	0.035
C_2	–0.035	1.000
Σl^2	1.001225	1.001225
$\sqrt{\Sigma l^2}$	1.000612	1.000612
$\sqrt{\Sigma l^2} = D_{23}$	0.9994	0.9994

(c) Direction Cosines (T_{23})

	A_3	B_3	C_3	Σl^2
A_2	1.000	0.0000	0.000	1.000
B_2	0.000	0.9994	0.035	1.000
C_2	0.000	–0.035	0.9994	1.000
Σl^2	1.000	1.000	1.000	1.000

(d) Third Rotated Factor Matrix (O_3)

	A_3	B_3	C_3	h^2
1.	0.714	–0.009	0.293	0.595
2.	0.748	0.016	0.219	0.607
3.	0.269	0.500	–0.337	0.435
4.	0.259	0.539	–0.330	0.465
5.	0.005	0.448	0.474	0.425
6.	–0.003	0.487	0.444	0.434

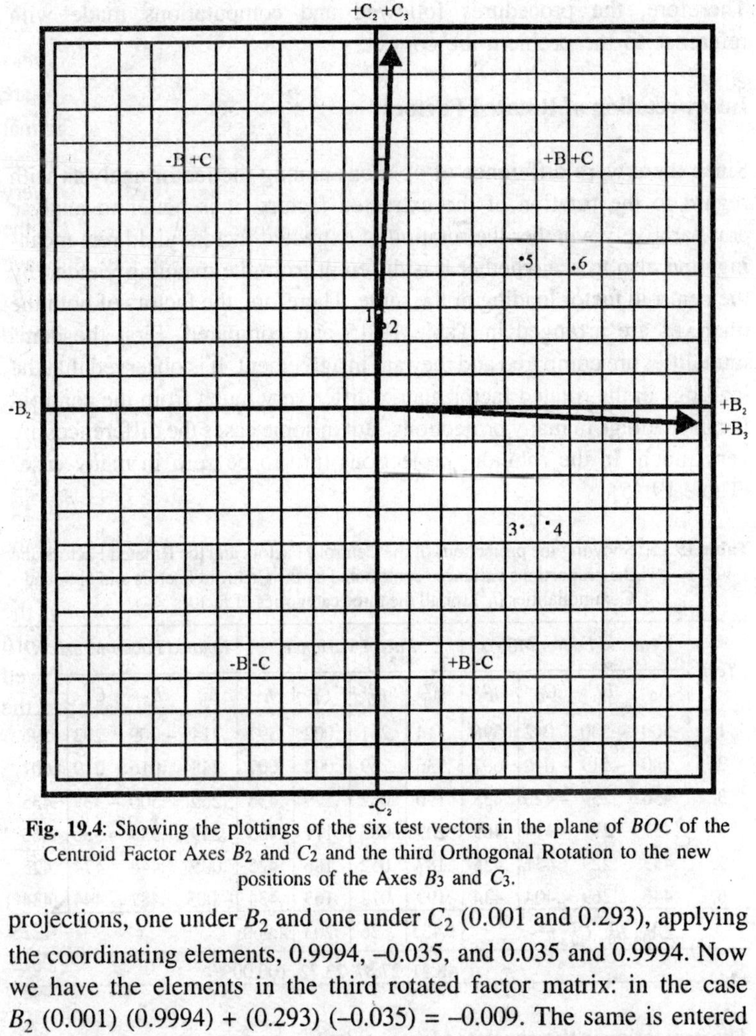

Fig. 19.4: Showing the plottings of the six test vectors in the plane of *BOC* of the Centroid Factor Axes B_2 and C_2 and the third Orthogonal Rotation to the new positions of the Axes B_3 and C_3.

projections, one under B_2 and one under C_2 (0.001 and 0.293), applying the coordinating elements, 0.9994, –0.035, and 0.035 and 0.9994. Now we have the elements in the third rotated factor matrix: in the case B_2 (0.001) (0.9994) + (0.293) (–0.035) = –0.009. The same is entered under B_3. In the case of C_2, (0.001) (0.035) + (0.293) (0.9994) = 0.293, which is entered under C_3. The rotated factor matrix O_3 is computed and formed in the same manner (Table 19.14).

In the last step the communalities (h^2) for the third rotated factor matrix and for the centroid factor matrix are worked out and checked. It is observed that the two communalities have almost the same values.

Therefore, the procedures followed and computations made with reference to the problem are correct.

Interpretation of Rotated Factors

Since there were difference of opinions among the factor analysts with regard to the rotation of the extracted factors, it is better to analyse comparatively whether the rotation of extracted factors yield real meaning, and also to see whether it is different from the meaning yielded by the centroid factor loading or the same. Therefore, the factors of both the matrixes are arranged in Table 19.15 and compared. First the communalities are compared and they are in agreement. It is observed that the loadings in the rotated factor matrix differ very much from the centroid factor loadings in many projections. But in some cases the differences are very much. In the rotation, projections turn to be zero in many cases (Table 19.15).

Table 19.15 : Showing the projections of the Centroid Factors and the Rotated Factors and the proportions variances contributed by the Centroid Factors and obtained Communalities (h^2) for all the three categories of factors

Test	*Centroid Factor Matrix*				*Factor Variance*				*Rotated Factor Matrix (O)*			
	A_0	B_0	C_0	h^2	A_0^2	B_0^2	C_0^2	h^2	A_3	B_3	C_3	h^2
1.	.560	–.530	.042	.596	.314	.281	.002	.597	.714	–.009	.293	.596
2.	.580	–.519	–.042	.607	.336	.269	.002	.607	.748	.016	.219	.607
3.	.436	.259	–.423	.435	.190	.067	.179	.436	.269	.500	–.337	.435
4.	.457	.291	–.414	.465	.209	.085	.171	.465	.259	.539	–.330	.465
5.	.433	.229	.431	.425	.187	.052	.186	.425	.005	.448	.474	.425
6.	.446	.269	.404	.434	.199	.072	.163	.434	–.003	.487	.444	.434
	$\Sigma A_0^2, B_0^2, C_0^2, h^2$				1.435	.826	.703	2.960				
					48.41	27.87	23.72	100.00%				

The projection in the unrotated factor matrix in certain cases are not giving us clear idea about the performance of the tests. But after rotation the same test projections show what the tests actually measure. There are highly significant loadings in the tests 1 and 2, like, 0.714 and 0.748. The centroid factor loadings also in these two tests have high loadings, like 0.560 and 0.580. But in rotation this is enhanced considerably. The first factor has exhausted 48.41 per cent of the total variance. Therefore, the first factor

A, may be identified as verbalization since the tests 1 and 2 are dealing with reading ability and verbal comprehension (Table 19.15).

Tests 3 and 4 have appreciable loadings in the second factor. Test 3, is dealing with arithmetic reasoning and test 4 is dealing with mechanical ability with the loadings of 0.500 and 0.539 respectively. The factor *B* has exhausted 27.87 per cent of the total variance. This factor *B*, has high loadings in tests 4 and 5 also with 0.448, and 0.487 respectively. The factor *B* appears to be a complex one, involving tests 3, 4, 5 and 6. Students who are proficient in mathematics wish to take up mechanical engineering course. Therefore, these two abilities may go together and also mechanical ability involve spatial relations and assembling of various parts. Hence, there is something in common in the last four tests. Therefore, the factor *B* may be stated as mechanical perception (Table 19.15).

Tests 5 and 6 have significant loadings in factor *C*. All the other four tests 1, 2, 3, and 4 have poor and negative loadings. Tests 5 and 6 have exhausted 23.72 per cent of the total variance with loadings of 0.474 and 0.444 respectively. These two tests are mainly concerned with spatial relations and assembling parts. In assembling task the perception of space and relating the parts to the space is very important. Hence, these two abilities may go together. Therefore, the factor *C* may be stated as spatial perception. It is very difficult to evolve this type meaning to the loadings of the factors in the centroid factors.

APPENDICES

Appendix -A

TABLE OF RANDOM NUMBERS

1	2	3	4	5	6	7	8	9	10	11	12	13	14	15	16	17	18	19	20	21	22	23	24	25
10	09	73	25	33	76	52	01	35	86	34	67	35	48	76	80	95	90	91	17	39	29	27	49	45
37	54	20	48	05	64	89	47	42	96	24	80	52	40	37	20	63	61	04	02	00	82	29	16	65
08	42	26	89	53	19	64	50	93	03	23	20	90	25	60	15	95	33	47	64	35	08	03	36	06
99	01	90	25	29	09	37	67	07	15	38	31	13	11	65	88	67	67	43	97	04	43	62	76	59
12	80	79	99	70	80	15	73	61	47	64	03	23	66	53	98	95	11	68	77	12	17	17	68	33
66	06	57	47	17	34	07	27	68	50	36	69	73	61	70	65	81	33	98	85	11	19	92	91	70
31	06	01	08	05	45	57	18	24	06	35	30	34	26	14	86	79	90	74	39	23	40	30	97	32
85	26	97	76	02	02	05	16	56	92	68	66	57	48	18	73	05	38	52	47	18	62	38	85	79
63	57	33	21	35	05	32	54	70	48	90	55	35	75	48	28	46	82	87	09	83	49	12	56	24
73	79	64	57	53	03	52	96	47	78	35	80	83	42	82	60	93	52	03	44	35	27	38	84	35
98	52	01	77	67	14	90	56	86	07	22	10	94	05	58	60	97	09	34	33	50	50	07	39	98
11	80	50	54	31	39	80	82	77	32	50	72	56	82	48	29	40	52	42	01	52	77	56	78	51
83	45	29	96	34	06	28	89	80	83	13	74	67	00	78	18	47	54	06	10	68	71	17	78	17
88	68	54	02	00	86	50	75	84	01	36	76	66	79	51	90	36	47	64	93	29	60	91	10	62
99	59	46	73	48	87	51	76	49	69	91	82	60	89	28	93	78	56	13	68	23	47	83	41	13
65	48	11	76	74	17	46	85	09	50	58	04	77	69	74	73	03	95	71	86	40	21	81	65	44
80	12	43	56	35	17	72	70	80	15	45	31	82	23	74	21	11	57	82	53	14	38	55	37	63
74	35	09	98	17	77	40	27	72	14	43	23	60	02	10	45	52	16	42	37	96	28	60	26	55
69	91	62	68	03	66	25	22	91	48	36	93	68	72	03	76	62	11	39	90	94	40	05	64	18
09	89	32	05	05	14	22	56	85	14	46	42	75	67	88	96	29	77	88	22	54	38	21	45	98

91	49	91	45	21	68	47	92	76	86	46	16	28	35	54	94	75	08	99	23	37	08	92	00	48
80	33	69	45	58	26	94	03	68	58	70	29	73	41	35	53	14	03	33	40	42	05	08	23	41
44	10	48	19	49	85	15	74	79	54	32	97	92	65	75	57	60	04	08	81	22	22	20	64	13
12	55	07	37	42	11	10	00	20	40	12	86	07	46	97	96	64	48	94	39	28	70	72	58	15
63	60	64	93	29	16	50	53	44	84	40	21	95	25	63	43	65	17	70	82	07	20	73	17	90
61	19	69	04	46	26	45	74	77	74	51	92	43	37	29	65	39	45	95	93	42	58	26	05	27
15	47	44	52	66	95	27	07	99	53	59	36	78	38	48	82	39	61	01	18	33	21	15	94	66
94	55	72	85	73	67	89	75	43	87	54	62	24	44	31	91	19	04	25	92	92	92	74	59	73
42	48	11	62	13	97	34	40	87	21	16	86	84	87	67	03	07	11	20	59	25	70	14	66	70
23	52	37	83	17	73	20	88	98	37	68	93	59	14	16	26	25	22	96	63	05	52	28	25	62
04	49	35	24	94	75	24	63	38	24	45	86	25	10	25	61	96	27	93	35	65	33	71	24	72
00	54	99	76	54	64	05	18	81	59	96	11	96	38	96	54	69	28	23	91	23	28	72	95	29
35	96	31	53	07	26	89	80	93	54	33	35	13	54	62	77	97	45	00	24	90	10	33	93	33
59	80	80	83	91	45	42	72	68	42	83	60	94	97	00	13	02	12	48	92	78	56	52	01	06
46	05	88	52	36	01	39	09	22	86	77	28	14	40	77	93	91	08	38	47	70	61	74	29	41
32	17	90	05	97	87	37	92	52	41	05	56	70	70	07	86	74	31	71	57	85	39	41	18	38
69	23	46	14	06	20	11	74	52	04	15	95	66	00	00	18	74	39	24	23	97	11	89	63	38
19	56	54	14	30	01	75	87	53	79	40	41	92	15	85	66	67	43	68	06	84	96	28	52	07
45	15	51	49	38	19	47	60	72	46	43	66	79	45	43	59	04	79	00	33	20	82	66	95	41
94	86	43	19	94	36	16	81	08	51	34	88	88	15	53	01	54	03	54	56	05	01	45	11	76

Reprinted by permission of Free Press, Glencoe, Ill., 1955, Abridged from the RAND Corporation, A Million Random Digits.

Appendix-B

TABLE OF 't' RATIO

Levels of Significance of 't' Ratio

df	0.4	0.3	0.2	0.1	0.05	0.02	0.01	0.001
1	1.376	1.963	3.078	6.314	12.706	31.821	63.657	636.619
2	1.061	1.386	1,886	2.920	4.303	6.965	9.925	31.598
3	0.978	1.250	1.638	2.353	3.182	4.541	5.841	12.941
4	0.941	1.190	1.533	2.132	2.776	3.747	4.604	8.610
5	0.920	1.156	1.476	2.015	2.571	3.365	4.032	6.859
6	0.906	1.134	1.440	1.943	2.447	3.143	3.707	5.959
7	0.896	1.119	1.415	1.895	2.365	2.998	3.499	4.405
8	0.889	1.108	1.397	1.860	2.306	2.896	3.355	5.041
9	0.883	1.100	1.383	1.833	2.262	2.821	3.250	4.781
10	0.879	1.093	1.372	1.812	2.228	2.764	3.169	4.587
11	0.876	1.088	1.363	1.796	2.201	2.718	3.106	4.437
12	0.873	1.083	1.356	1.782	2.179	2.681	3.055	4.318
13	0.870	1.079	1.350	1.771	2.160	2.650	3.012	4.221
14	0.868	1.076	1.345	1.761	2.145	2.624	2.977	4.140
15	0.866	1.074	1.341	1.753	2.131	2.602	2.947	4.073
16	0.865	1.071	1.337	1.746	2.120	2.583	2.921	4.015
17	0.863	1.069	1.333	1.740	2.110	2.567	2.898	3.965
18	0.862	1.067	1.330	1.737	2.101	2.552	2.878	3.922
19	0.861	1.066	1.328	1.729	2.093	2.539	2.861	3.883
20	0.860	1.064	1.325	1.725	2.086	2.528	2.845	3.850
21	0.859	1.063	1.323	1.721	2.080	2.518	2.831	3.819
22	0.858	1.061	1.321	1.717	2.074	3.792	2.819	3.792
23	0.858	1.060	1.319	1.714	2.069	2.500	2.807	3.767
24	0.857	1.059	1.318	1.711	2.064	2.492	2.797	3.745
25	0.856	1.058	1.316	1.708	2.060	2.485	2.787	3.725
26	0.856	1.058	1.315	1.706	2.056	2.479	2.779	3.707
27	0.855	1.057	1.314	1.703	2.052	2.473	2.771	3.690
28	0.855	1.056	1.313	1.701	2.048	2.467	2.763	3.674
29	0.854	1.055	1.311	1.699	2.045	2.462	2.756	3.659
30	0.854	1.055	1.310	1.697	2.042	2.457	2.750	3.646
∞	0.842	1.036	1.282	1.645	1.960	2.326	2.576	3.291

Reprinted by permission of Addison Wesley Longmen Ltd., Abridged from IV of Fisher and Yates : Statistical Tables for Biological, Agricultural and Medical Research (0-582-44525-6).

Appendix-C

TABLE OF *F* RATIO

P_2 \ P_1	P	1	2	3	4	5	6	7	8	9	10	12	15	20	24	30	∞
		Degrees of Freedom Associated with Numerator															
1	0.05	161	200	216	225	230	234	237	239	241	242	244	246	248	249	250	254
	0.01	4,052	4,999	5,403	5,625	5,764	5,859	5,928	5,981	6,022	6,056	6.106	6,157	6,200	6,235	6,261	6,366
2	0.05	18.51	19.00	19.16	19.25	19.30	19.33	19.36	19.37	19.38	19.39	19.41	19.43	19.45	19.45	19.46	19.50
	0.01	98.49	99.00	99.17	99.25	99.30	99.33	99.34	99.36	99.38	99.40	99.42	99.43	99.45	99.46	99.47	99.50
3	0.05	10.13	9.55	9.28	9.12	9.01	8.94	8.88	8.84	8.81	8.78	8.74	8.70	8.65	8.64	8.62	8.53
	0.01	34.12	30.82	29.46	28.71	28.24	27.91	27.67	27.49	27.34	27.23	27.05	26.87	28.69	26.60	26.50	26.13
	0.001	167.00	148.50	141.10	137.10	134.60	132.20	132.00	130.60	130.00	123.00	128.30	127.00	126.00	125.70	125.00	123.50
4	0.05	7.71	6.94	6.59	6.39	6.26	6.16	6.09	6,04	6.00	5.96	5.91	5.86	5.80	5.77	5.75	5.63
	0.01	21,20	18.00	16.69	15.98	15.52	15.21	14.98	14.80	14.66	14.54	14.37	14.20	14.08	13.96	13.84	13.46
	0.001	74.14	61.25	56.18	53.44	51.71	50.53	49.70	49.00	48.50	48.00	47.41	46.80	46.10	45.57	45.40	44.05
5	0.05	6.61	5.79	5.41	5.19	5.05	4.95	4.88	4.82	4.78	4.74	4.68	4.62	4.56	4.53	4.50	4.36
	0.01	16.26	13.27	12.06	11.39	10.97	10.67	10.45	10.27	10.15	10.05	9.89	9.72	9.55	9.47	9.38	9.02
	0.001	47.18	37.12	33.20	31.09	29.75	28.84	28.20	27.64	27.20	26.90	26.42	25.90	25.40	25.14	24.90	23.79
6	0.05	5.99	5.14	4.76	4.53	4.39	4.28	4.21	4,15	4.10	4.06	4.00	3.94	3.87	3.84	3.81	3.67
	0.01	13.74	10.92	9.78	9.15	8.75	8.47	8.26	8.10	7.98	7.87	7.72	7.56	7.40	7.31	7.23	6.88
	0.001	35.50	27.00	23.20	21.93	20.81	20.03	19.50	19.03	18.70	18.00	17.97	17.60	17.10	16.85	16.70	15.75
7	0.05	5.59	4.74	4.35	4.12	3.97	3.87	3.79	3.73	3.68	3.63	3.57	3.51	3.44	3.41	3.38	3.23
	0.01	12.25	9.55	8.45	7.85	7.46	7.19	7.00	6.84	6.71	6.62	6.47	6.31	6.16	6.07	5.99	5.65
	0.001	29.25	27.00	18.77	17.19	16.71	15.52	15.00	14.63	14.30	14.10	13.71	13.30	12.90	12.73	12.50	11.70
8	0.05	5.32	4.46	4.07	3.84	3.69	3.58	3.50	3,44	3.39	3.34	3.28	3.22	3.15	3.12	3.08	2.93
	0.01	11.25	8.65	7.59	7.01	6.63	6.37	6.19	6.03	5.91	5.82	5.67	5.52	5.36	5.28	5.20	4.86
	0.001	25.42	21.69	15.83	14.39	13.49	12.86	12.40	12.04	11.80	11.50	11.19	10.80	10.50	10.30	10.10	9.33

Degrees of Freedom Associated with Denominator

P_2 \ P_1	P	1	2	3	4	5	6	7	8	9	10	12	15	20	24	30	∞
		Degrees of Freedom Associated with Numerator															
9	0.05	5.12	4.26	3.86	3.63	3.48	3.37	3.29	3.23	3.18	3.13	3.07	3.01	2.94	2.90	2.86	2.71
	0.01	10.56	8.02	6.99	6.42	6.06	5.80	5.62	5.47	5.35	5.26	5.11	4.96	4.81	4.73	4.65	4.31
	0.001	22.86	18.49	13.90	12.56	11.71	11.13	10.70	10.37	10.10	9.89	9.59	9.24	8.90	8.72	8.55	7.81
10	0.05	4.96	4,10	3.71	3.48	3.33	3.22	3.14	3.07	3.02	2.97	2.91	2.85	2.77	2.74	2.70	2.54
	0.01	10.04	7.56	6.55	5.99	5.64	5.39	5.21	5.06	4.95	4.85	4.71	4.56	4.41	4.33	4.25	3.91
	0.001	21.00	16.39	12.35	11.28	10.48	9.92	9.52	9.20	8.96	8.75	8.45	8.13	7.80	7.64	7.47	6.76
11	0.05	4.48	3.98	3.59	3.46	3.20	3.09	3.01	2.95	2.90	2.86	2.79	2.72	2.65	2.61	2.57	2.40
	0.01	9.65	7.20	6.22	5.67	5.32	5.07	4.88	4.74	4.62	4.54	4.40	4.25	4.10	4.02	3.94	3.00
	0.001	19.69	14.91	11.56	10.35	9.58	9.05	8.66	8.35	8.12	7.92	7.63	7.32	7.01	6.85	6.68	6.00
12	0.05	4.75	3.88	3.49	3.26	3.11	3.00	2.92	2.85	2.80	2.76	2.69	2.62	2.54	2.51	2.47	2.30
	0.01	9.33	6.93	5.95	5.41	5.06	4.82	4.65	4.50	4.39	4.30	4.16	4.01	3.86	3.78	3.70	3.36
	0.001	18.64	13.89	10.80	9.63	8.89	8.38	8.00	7.71	7.48	7.29	7.00	6.79	6.40	6.25	6.09	5.42
13	0.05	4.67	3.80	3.41	3.18	3.02	2.92	2.84	2.77	2.72	2.67	2.60	2.53	2.46	2.42	2.38	2.21
	0.01	9.07	6.70	5.74	5.20	4.86	4.62	4.44	4.30	4.19	4.10	3.96	3.82	3.66	3.59	3.51	3.17
	0.001	17.81	12.97	10.21	9.07	8.35	7.86	7.49	7.21	6.98	6.80	6.52	6.23	6.13	5.78	5.63	4.97
14	0.05	4.60	3.74	3.34	3.11	2.96	2.85	2.77	2.70	2.63	2.60	2.53	2.46	2.39	2.35	2.31	2.13
	0.01	8.86	6.51	5.56	5.03	4.69	4.46	4.28	4.14	4.03	3.94	3.80	3.66	3.51	3.43	3.35	3.00
	0.001	17.14	12.31	9.73	8.62	7.92	7.43	7.08	6.80	6.58	6.40	6.13	5.85	5.56	5.41	5.25	4.60
15	0.05	4.54	3.68	3.29	3.06	2.90	2.79	2.70	2.64	2.59	2.55	2.48	2.40	2.33	2.29	2.25	2.07
	0.01	8.68	6.36	5.42	4.89	4.56	4.32	4.14	4.00	3.89	3.80	3.67	3.52	3.37	3.29	3.21	2.87
	0.001	16.59	11.78	9.34	8.25	7.57	7.09	6.74	6.47	6.26	6.08	5.81	5.54	5.25	5.10	4.95	4.31
20	0.05	4.35	3.49	3.10	2.87	2.71	2.60	2.52	2.45	2.40	2.35	2.28	2.20	2.12	2.08	2.04	1.84
	0.01	8.10	5.85	4.94	4.43	4.10	3.87	3.71	3.56	3.45	3.37	3.23	3.09	2.94	2.86	2.78	2.42
	0.001	14.82	9.95	8.10	7.10	6.46	6.02	5.69	5.44	5.24	5.08	4.82	4.56	4.29	4.15	4.00	3.38

Degrees of Freedom Associated with Denominator

P_1 / P_2	P	1	2	3	4	5	6	7	8	9	10	12	15	20	24	30	∞
		Degrees of Freedom Associated with Numerator															
24	0.05	4.26	3.40	3.01	2.78	2.62	2.51	2.42	2.36	2.30	2.25	2.18	2.11	2.03	1.98	1.94	1.73
	0.01	7.82	5.61	4.72	4.22	3.90	3.67	3.50	3.36	3.26	3.17	3.03	2.89	2.74	2.66	2.58	2.21
	0.001	14.00	9.39	7.55	6.59	5.98	5.55	5.23	4.99	4.80	4.54	4.39	4.14	3.87	3.74	3.59	2.97
26	0.05	4.23	3.37	2.98	2.74	2.59	2.47	2.39	2.32	2.27	2.22	2.15	2.07	1.99	1.95	1.90	1.69
	0.01	7.72	5.53	4.64	4.14	3.82	3.59	3.42	3.29	3.18	3.09	2.96	2.81	2.66	2.58	2.50	2.13
	0.001	13.70	8.12	7.36	6.41	5.80	5.38	5.07	4.83	4.64	4.48	4.24	3.99	3.72	3.59	3.43	2.82
30	0.05	4.17	3.32	2.92	2.69	2.53	2.42	2.34	2.27	2.21	2.16	2.09	2.01	1.93	1.89	1.84	1.62
	0.01	7.56	5.39	4.51	4.02	3.70	3.47	3.30	3.17	3.06	2.98	2.84	2.70	2.55	2.47	2.39	2.01
	0.001	13.29	9.97	7.05	6.12	5.53	5.12	4.82	4.58	4.39	4.24	4.00	3.75	3.49	3.36	3.22	2.59
40	0.05	4.08	3.23	2.84	2.61	2.45	2.34	2.25	2.18	2.12	2.07	2.00	1.92	1.84	1.79	1.74	1.51
	0.01	7.31	5.18	4.31	3.83	3.51	3.29	3.12	2.99	2.88	2.80	2.66	2.52	2.37	2.29	2.20	1.80
	0.001	12.61	8.23	6.60	5.70	5.13	4.73	4.44	4.21	4.02	3.87	3.64	3.40	3.15	3.01	2.87	2.23
60	0.05	4.00	3.15	2.76	2.52	2.37	2.25	2.17	2.10	2.04	1.99	1.92	1.84	1.75	1.70	1.65	1.39
	0.01	7.08	4.98	4.13	3.65	3.34	3.12	2.95	2.82	2.72	2.63	2.50	2.35	2.20	2.12	2.03	1.60
	0.001	11.97	7.76	6.17	5.31	4.76	4.37	4.09	3.87	3.69	3.54	3.30	3.08	2.83	2.69	2.55	1.89
120	0.05	3.92	3.07	2.68	2.45	2.29	2.17	2.09	2.02	1.96	1.91	1.83	1.75	1.66	1.61	1.55	1.25
	0.01	6.83	4.79	3.95	3.48	3.17	2.96	2.79	2.66	2.56	2.47	2.34	2.19	2.03	1.95	1.86	1.38
	0.001	11.40	7.32	5.79	4.95	4.42	4.04	3.77	3.55	3.38	3.24	3.02	2.78	2.53	2.48	2.26	1.54
∞	0.05	3.84	3.00	2.60	2.37	2.21	2.10	2.01	1.94	1.88	1.83	1.75	1.67	1.57	1.52	1.46	1.00
	0.01	6.63	4.61	3.78	3.32	3.02	2.80	2.64	2.51	2.41	2.32	2.18	2.04	1.88	1.79	1.70	1.00
	0.001	10.83	6.91	5.42	4.62	4.10	3.74	3.47	3.27	3.10	2.96	2.74	2.51	2.27	2.13	1.99	1.00

Degrees of Freedom Associated with Denominator

Reprinted by permission of Addison Wesley Longman Ltd.,
Abridged from Table V of Fisher, and Yates : Statistical Tables for Biological, Agricult ural and Medical Research (0-582-44525-6).

BIBLIOGRAPHY

Anastasi, A. *Psychological Testing*. 2nd ed. New York : Macmillan, 1961.

Black, J.A., and Champion, D.J. *Methods and Issues in Social Research*, New York : John Wiley & Sons, Inc, 1976.

Blalock, H.M. Jr. *Social Statistics*. New York : McGraw- Hill, 1960.

Bogardus, E.S. Social Distance and Its Practical Implications, *Sociology and Social Research*, Vol. *XVII*, 1933, pp. 265-71.

Burt, C. *The Factors of the Mind*, New York : Macmillan, 1941.

Burt, C., and Watson, H. Factor Analysis of Assessment of a Single Person. *British Journal of Psychological Statistics*, 1951, **4**, 179-91.

Candland, D.K. *Psychology : The Experimental Approach*. New York : McGraw-Hill Book Company, 1968.

Cattell, R.B. *Description and Measurement of Personality*. Yonkers, N.Y. : World, 1941.

Cattell, R.B., Saunders, D.R. and Stice, G. *Handbook for the Sixteen Personality Factor Questionnaire. Test. Forms A and B. Campaign, III* : Institute for Personality and Ability Testing, 1950.

Cattell, A.K.S., and Rhymer, R.M. P-techniques demonstrated in determining Psychophysical sources traits in a normal individual. *Psychometrika*, 1947, **12**, 217-88.

Cohan, Morris and Nagel, E. *An Introduction to Logic and Scientific Method*. New York : Harcourt Brace, 1934.

Coleman, J.S., Katz, E and Menzel, N.M. Diffusion of Innovation among Physicians, *Sociometry*, 1957, **20**, 253- 70.

Coleman, J.S. Rational Analysis : The Study of Social Organizations with Methods, *Human Organization*, **17**, 1958-59, 28-36.

Cronbach, L.J. Test reliability : Its meaning and determination, *Psychometrika*, 1947, **12**, 1-16.

Cronbach, L.J. *Essentials of Psychological Testing* (2nd. ed.) New York : Harper and Row, 1966.

Cureton, E.E. "*Validity*" *in Educational Measurement*. Ed. E.F. Lindquist, Washington, DC : American Council on Education, 1951.

D'Amato, M.R. *Experimental Psychology : Methodology, Psychophysics and Learning*. (Inter Student ed.). New York : McGraw-Hill Book Company, 1970.

Duncan, D.B. Multiple range and multiple F-test. *Biometrika*, 1955, 11, 1-42.

Ebel, R.L. *Measuring Educational Achievement*, New Delhi : Prentice-Hall of India Private Ltd., 1966.

Edwards, A.L. *Statistical Methods for Behavioural Sciences*, New York : Holt, Rinehart & Winston, 1961.

Edwards, A.J., and Scannell, D.P. *Educational Psychology* (Indian Edition), Bombay : Allied Publishers Private Limited, 1968.

Epstein, W. and Shontz, F. *Psychology in Progress*. New York : Holt Rinehart and Winston, Inc., 1970.

Fruchter, . *Introduction to Factor Analysis*. Prince, New Jersey: D. Van Nostrand Company, Inc., 1954.

Guilford, J.P. *Psychometric Methods*, (2nd ed.) Bombay; New Delhi: Tata-McGraw-Hill Publishing Co. Ltd., 1954.

Guilford, J.P. and Fruchter, B. *Fundamental Statistics in Psychology and Education*, (6th ed.). Singapore : McGraw- Hill Book Company, 1987.

Gullikson, H. *Theory of Mental Testing*. New York: Wiley, 1950.

Hagood, Margaret Jarman. *Statistics for Sociologists*. New York: Holt, 1941.

Hathaway, S.R. and McKinley, J.C. *The Minnesota Multiphasic Personality Inventory*. Mineapolis : University of Minnesota Press, 1943.

Hoel, P.G. *Introduction to Mathematical Statistics*. New York: Wiley, 1947.

Holzinger, K.J. and Harman, H.H. *Factor Analysis: A Synthesis of Factorial Methods*. Chicago: University of Chicago Press, 1941.

Hotelling, H. Analysis of a complex statistical variables into principal components. *Journal of Educational Psychology*, 1933, **24**, 417-41.

Huddleston, E.M. Test Development on the basis of content validity; *Educational and Psychological Measurement XVI* (Autumn) 1956, 283-93.

Johnson, P.O. *Statistical Methods in Research*. New York: Prentice-Hall, 1949.

Kelley, T.C. *Essential Traits of Mental Life*. Cambridge Mass: Harvard University Press, 1935.

Kuder, B.I. and Richardson, M.W. The Theory of Estimation of Test Reliability, *Psychometrika*, II (September), 1937, 151-60.

Lindquist, E.F. *Educational Measurement*. Washington, D.C.: American Council of Education, 1951.

Lindvall, C.M. *Testing and Evaluation : An Introduction*. New York: Harcourt Brace & World, Inc, 1961.

Lyman, H.B. *Test Scores and What they mean*. Englewoodcliffs, N.J.: Prentice-Hall, 1963.

Magnussun, D. *Test Theory: Reading*. Mass: Addison- Wesley, 1967.

McGuigan, F.J. *Experimental Psychology*. 3rd Edition. Englewood Cliffs, New Jersey: Prentice-Hall, INC. 1978.

Mood, A.M. *Introduction to the Theory of Statistics*. New York: McGraw-Hill, 1950.

Moskowitz, M.J. and Orgel, A.R. *General Psychology*. Boston: Houghton Mifflin Company, 1969.

Peatman, J.G. *Descriptive and Sampling Statistics*. New York: Harper and Brothers, 1947.

Quinn McNemar, *Psychological Statistics*. New York: John Wiley and Sons, Inc., 1955.

Reyburn, M.A., and Taylor, J.G. On the Interpretation of Common factors, *Psychometrika*, 1946, **11**, 51-56.

Ryan, T.A. Multiple comparisons in Psychological Research, *Psychological Bulletin*, 1959, **56** 26-47.

Scheffe, H. A method of judging all contrasts in the Analysis of Variance, *Biometrika*, 1953, **40**, 87-104.

Selltiz, C. *et al*. *Research Methods in Social Sciences*. New York: Holt Rinehart and Winston, 1961.

Sheridan, C.L. *Methods in Experimental Psychology*. New York: Hold Rinehart and Winston, 1970.

Spearman, C. "General Intelligence" objectively determined and measured. *American Journal of Psychology*, 1904, **15**, 201-93.

Spearman, C. *Abilities of Man*, New York: Macmillan, 1927.

Technical Recommendations for Psychological Tests and Diagnostic Techniques. *Psychological Bulletin*, 1954, **51**, Supplement.

Thomas, L.G. Mental Tests as Instrument of Science. *Psychological Monograph*, 1942, **54**, No. 245.

Thompson, G.H. *The Factorial Analysis of Human Ability*. 3rd ed. Boston: Houghton Mufflin, 1948.

Thorndike, R.L. and Hagan, E. *Measurement and Evaluation in Psychology and Education*. (2nd ed.). New Delhi: Wiley Eastern Private Ltd., 1961.

Thurstone, L.L. *Vectors of Mind*. Chicago: University of Chicago Press, 1935.

Thurstone, L.L. Primary Mental Abilities. *Psychometric Monograph*, Chicago, Illinois: University of Chicago Press, 1940.

Thurstone, L.L. *Multiple Factor Analysis*. Chicago: University of Chicago Press, 1947.

Wundt, W. *Grundriss der Psychologie*, 1896. Translated by C.H. Judd. *Outlines of Psychology*. Leipzig: Englemann, 1897.

Zimmerman, W.S. A simple graphical method for orthogonal rotation of axes. *Psychometrika*, 1946, **11**, 51-56.

INDEX